SpringerBriefs in Biological Imaging

Series Editor
Douglas J. Taatjes

For further volumes:
http://www.springer.com/series/10359

Bhanu P. Jena

NanoCellBiology
of Secretion

Imaging Its Cellular and Molecular
Underpinnings

Foreword by Lloyd L. Anderson, Ph.D., D.Sc.

 Springer

Bhanu P. Jena
Director, NanoBioScience Institute
Department of Physiology
Wayne State University School of Medicine
Detroit, MI 48201, USA
bjena@med.wayne.edu

ISSN 2193-3359   e-ISSN 2193-3367
ISBN 978-1-4614-2437-6   e-ISBN 978-1-4614-2438-3
DOI 10.1007/978-1-4614-2438-3
Springer New York Dordrecht Heidelberg London

Library of Congress Control Number: 2011945438

© Springer Science+Business Media, LLC 2012
All rights reserved. This work may not be translated or copied in whole or in part without the written
permission of the publisher (Springer Science+Business Media, LLC, 233 Spring Street, New York,
NY 10013, USA), except for brief excerpts in connection with reviews or scholarly analysis. Use in
connection with any form of information storage and retrieval, electronic adaptation, computer software,
or by similar or dissimilar methodology now known or hereafter developed is forbidden.
The use in this publication of trade names, trademarks, service marks, and similar terms, even if they
are not identified as such, is not to be taken as an expression of opinion as to whether or not they are
subject to proprietary rights.

Printed on acid-free paper

Springer is part of Springer Science+Business Media (www.springer.com)

Dedicated to my loving parents Manju Prova and Prafulla Kumar Jena

Bhanu P. Jena

Foreword

Eukaryotic cells from yeast to humans encode 6,000 to 20,500 proteins, and >1,000 proteins, i.e., a minimum of 5% of all coded proteins have been identified to be secreted in humans (*Genome Res* 2003, 13:2265–2270). While there has been much focus on protein biosynthesis (discovery of transcription, translation, the signal peptide, the ribosome and its structure–function), and their folding or misfolding leading to degradation (the prion protein, the proteasome), protein secretion following synthesis, and the release of various molecules from the cell essential for maintaining physiological homeostasis within organisms, has received little attention.

Critical observations made in the late 1940s by George E. Palade led him to state in his 1974 Nobel Lecture: "a fibrillar shell often appears around discharging zymogen granules when their membrane is already in continuity with the plasmalemma. It is continuous with the terminal web, it may consist of contractile proteins (actin? myosin?), and it may promote the expulsion of the secretion granule content." After almost 40 years, the above statement has proven to be indeed true, with discovery of the "porosome"—the universal secretory portal in cells by Bhanu P. Jena nearly 15 years ago, first in acinar cells of the exocrine pancreas (*Proc Natl Acad Sci USA* 1997, 94:316–321; *Cell Biol Int* 2002, 26:35–42; *Biochemistry* 2009, 49:4009–4018), and in chromaffin cells (*Ann New York Acad Sci* 2002, 971:254–256). Subsequently, in collaboration with Prof. Jena and his research team, we were able to confirm and establish the presence of porosomes in growth hormone cells of the pituitary gland (*Endocrinology* 2002, 143:1144–1148). In 2004, the porosome at the nerve terminal in neurons was discovered (*Cell Biol Int* 2004, 28:699–708; *J Microsc* 2008, 232:106–111), and similar structures identified using EM tomography (*J Neurosci* 2007, 27:6868–6877). Further studies made in the past decade have progressed our understanding of porosomes in the exocrine pancreas (*Biophys J* 2003, 84:1–7 and 85:2035–2043; *J Biomed Nanotechnol* 2007, 3:218–222; *J American Sci* 2011, 7:835–843), in astrocytes (*J Cell Mol Med* 2009, 13:365–372; *Cell Biol Int* 2009, 33:224–229), in β-cells of the endocrine pancreas (*J Cell Mol Med* 2004, 8:1–21; *Exp Biol Med* 2005, 230:307–319), in hair cells of the inner ear (*Cell Biol Int Rep* 2011, CNR20110005), and in RBL-2H3 and BMMC cells (*J Phys Chem B* 2010, 114:5971–5982). The pioneering discovery of the porosome as the universal secretory nanomachine in cells provides for the first time a molecular understanding of cell secretion, resulting in a paradigm shift in our understanding of the process.

Prof. Jena's studies briefly outlined in this book demonstrate that following a secretory stimulus, membrane-bound secretory vesicles transiently dock and fuse at the base of porosomes present at the cell plasma membrane to release intravesicular contents. Prof. Jena's work has further resulted in the determination of the molecular assembly and disassembly of membrane-associated t-/v-SNARE complex involved in membrane fusion, and determination of the molecular mechanism of vesicle swelling, and its requirement in the regulated release of intravesicular contents during cell secretion. Consequently, for the first time, these seminal discoveries explain the generation of partially empty secretory vesicles in the cytosol, observed following cell secretion.

Ames, IA, USA   Lloyd L. Anderson, Ph.D., D.Sc.

Preface

The unit of life the cell is encapsulated within a lipoprotein membrane bilayer that precisely regulates the entry and exit of ions and molecules into and out of the cell, essential to maintaining the delicate chemical balance required to sustain life. Certain molecules such as neurotransmitters, hormones, or digestive enzymes, synthesized within the cell, are packaged and stored within membrane-bound vesicular compartments, ready for release (secretion) outside the cell on demand. In eukaryotic cells, various physiological processes rely and/or are influenced and regulated by secretion that occurs in every organism, from the simple yeast to humans. For example, secretion of neurotransmitters at the nerve terminal enables neurotransmission, allowing thought, movement, and coordination. Similarly after a meal, secretion of digestive enzymes from the exocrine pancreas helps digest food. The consequent elevation of blood glucose following digestion triggers secretion of insulin from β-cells of the endocrine pancreas. Similarly, exposure to certain types of pollen, or to a parasite, elicits an allergic inflammatory immune response, stimulating mast cells to secrete histamine and other compounds. From these few examples, it becomes clear how critical cell secretion is to life, and the severe health consequence that impairment of the process may lead to, such as the various neurological, digestive, hormonal, or immune disorders, to name just a few.

Examination of cells following secretion using electron microscopy demonstrates increased presence of partially empty vesicles. This suggested that during the secretory process, only a portion of the vesicular contents are able to exit the cell. This could only be possible if the vesicle were to temporarily establish continuity with the cell plasma membrane, expel a portion of its contents, then detach, reseal, and withdraw into the cytosol (endocytose). In this way, the secretory vesicle could be reused for subsequent rounds of exo-endocytosis, until completely empty of its contents. In case of synaptic vesicles however, the empty vesicle can rapidly undergo refilling with neurotransmitters from the cytosol via the neurotransmitter transporters present at the synaptic vesicle membrane.

Up until 15 years ago, it was believed that during cell secretion, membrane-bound secretory vesicles completely merge at the cell plasma membrane resulting in the diffusion of intravesicular contents to the cell exterior and the compensatory retrieval of the excess membrane by endocytosis. This explanation however failed to explain the generation of partially empty vesicles observed in electron micrographs following secretion. Logically therefore, in a 1993 *News and Views* article in the journal *Nature*, Prof. Erwin Neher wrote "It seems terribly wasteful that, during the release of hormones and neurotransmitters from a cell, the membrane of a vesicle should merge with the plasma membrane to be retrieved for recycling only seconds or minutes later." The discovery of the universal and permanent secretory portals or nanomachines at the cell plasma membrane called *porosomes*, where membrane-bound secretory vesicles transiently dock and fuse to release intravesicular contents to the cell exterior, has finally resolved this conundrum. Further determination of the composition of porosomes, their structure and dynamics using high-resolution atomic force and electron microscopy, and their functional reconstitution into artificial lipid membrane provides a molecular understanding of the secretory process in cells. In agreement, it has been demonstrated that "secretory granules are recaptured largely intact following stimulated exocytosis in cultured endocrine cells" (*Proc Natl Acad Sci* 2003, 100:2070–2075); "single synaptic vesicles fuse transiently and successively without loss of identity" (*Nature* 2003, 423:643–647); and "zymogen granule exocytosis is characterized by long fusion pore openings and preservation of vesicle lipid identity" (*Proc Natl Acad Sci* 2004, 101:6774–6779). It made no sense all these years to contend that mammalian cells possess an "all or none" mechanism of cell secretion resulting from complete vesicle merger at the cell plasma membrane, when even single-cell organisms have developed specialized and sophisticated secretory machinery, such as the secretion apparatus of *Toxoplasma gondii*, contractile vacuoles in paramecium, and different types of secretory structures in bacteria. The discovery of the porosome, its structure, function, composition, and functional reconstitution in artificial lipid membrane has resulted in a paradigm shift in our understanding of the secretory process in cells.

In this *Springer Brief*, the discovery of the porosome—*the universal secretory nanomachine*, the fusion of membrane-bound secretory vesicles at the porosome base, and the molecular mechanism of secretory vesicle swelling and its requirement for content expulsion during cell secretion, is briefly discussed.

Detroit, MI, USA   Bhanu P. Jena, Ph.D.

Contents

NanoCellBiology of Secretion

Abstract Cells synthesize, store, and secrete, on demand, products such as hormones, growth factors, neurotransmitters, or digestive enzymes. Cellular cargos destined for secretion are packaged and stored in membranous sacs or vesicles, which are transported via microtubule and actin railroad systems, to dock and establish continuity at the base of specialized cup-shaped plasma membrane structures called *porosomes* to release their contents. The fusion of membrane-bound secretory vesicles at the porosome base, and the molecular mechanism of secretory vesicle swelling required for content expulsion during cell secretion, is discussed in this *Springer Brief*. These new findings in the last 15 years have resulted in a paradigm shift in our understanding of the secretory process in cells and has given birth to a new field in biology—*NanoCellBiology*.

Introduction

In the past 50 years, it was believed that during cell secretion, membrane-bound secretory vesicles completely merge (the complete flattening of the secretory vesicle membrane at the cell plasma membrane) at the cell plasma membrane resulting in the diffusion of intravesicular contents to the cell exterior and the compensatory retrieval of the excess membrane by endocytosis. This explanation of cellular secretion however failed to account for the generation of partially empty vesicles observed in electron micrographs following secretion and remained a major conundrum in the field. Furthermore, such "all or none" mechanism of cell secretion by complete merger of the secretory vesicle membrane at the cell plasma membrane leaves little regulation and control by the cell on the amount of content release during the secretory process. It made no sense for mammalian cells to possess such "all or none" mechanism of cell secretion, when, by contrast, even single-cell organisms have developed specialized and sophisticated secretory machinery, such as the secretion apparatus of *Toxoplasma gondii*, the contractile vacuoles in paramecium, or the various types of secretory structures in bacteria. Therefore, not surprising in the 1960s, experimental data concerning neurotransmitter release mechanisms by Katz and Folkow [1, 2] proposed that limitation of the quantal packet may be set by the nerve membrane, in which case the size of the packet released would correspond to a fraction of the vesicle content [3, 4]. Again in 1993 in a *News and Views* article in the journal *Nature* [5], Neher appropriately noted "It seems terribly wasteful that, during the release of hormones and neurotransmitters from a cell, the membrane of a vesicle should merge with the plasma membrane to be retrieved for recycling only seconds or minutes later." This conundrum regarding the molecular underpinning of cell secretion was finally resolved in 1996 following discovery of the *porosome*, the universal secretory portal in cells [6]. Porosomes are supramolecular cup-shaped lipoprotein structures at the cell plasma membrane, where membrane-bound secretory vesicles transiently dock and fuse to release intravesicular contents to the outside during cell secretion. In the past 15 years, the composition of the porosome, its structure and dynamics at nanometer resolution and in real time, and its functional reconstitution into artificial lipid membrane have been elucidated [6–43], providing a molecular understanding of the secretory process in cells. Since porosomes in exocrine and neuroendocrine cells measure 100–180 nm, and only 20–45% increase in porosome diameter is demonstrated following the docking and fusion of 0.2–1.2 μm in diameter secretory vesicles, it is apparent that secretory vesicles "transiently" dock and establish continuity, as opposed to complete merger at the porosome base to release intravesicular contents to the outside. In agreement, it is demonstrated that "secretory granules are recaptured largely intact after stimulated exocytosis in cultured endocrine cells" [44]; that "single synaptic vesicles fuse transiently and successively without loss of identity" [45]; and that "zymogen granule (the secretory vesicle in exocrine pancreas) exocytosis is characterized by long fusion pore openings and preservation of vesicle lipid identity" [46].

Microtubules have been recognized as the railroad for movement of organelles over long distances within the cell (>1 μm), whereas the actin railroad system is responsible for transport over shorter distances, typically from tens to a few hundred nanometers. Thus, microtubule-dependent motors such as kinesin and kinesin-related proteins, and the superfamily of actin-dependent myosin motors, have all been implicated in intracellular organelle transport [47, 48]. Myosin motors include the conventional myosin (myosin II) and a large group of unconventional myosins (myosin I, III, V, and VI). In recent years, the prime candidate for secretory vesicle transport in cells has been reported to be the class V of myosin motors [49–51]. Myosin V is composed of two heavy chains that dimerize via a coiled-coil motif located in the stalk region of the heavy chain [52]. The heavy chain contains an amino-terminal actin-binding motor domain [52], followed by a neck region where up to six regulatory light chains can bind. The carboxy-terminus globular domain of the heavy chain is thought to mediate organelle-binding specificity [53]. Interaction between the actin and the microtubule transport system seems to be a requirement for the correct delivery of intracellular cargos such as secretory vesicles [54–56]. Studies have been undertaken to determine whether secretory vesicles in live cells remain free floating, only to associate with the transport systems following a secretory stimulus, or whether they always remain tethered in the cell. Studies using isolated live pancreatic acinar cells demonstrate that all secretory vesicles within cells remain tethered and are not free floating [57]. Nocodazole and cytochalasin B disrupt much of this tether. Immunoblot analysis of isolated secretory vesicles further demonstrates the association of actin, myosin V, and kinesin with them [57]. These studies reveal for the first time that secretory vesicles in live pancreatic acinar cells are tethered and not free floating, suggesting that following vesicle biogenesis, they are placed on their defined railroad track, ready to be transported to their final destination when required [57]. This makes sense, since precision and regulation are the hallmark of all cellular process and therefore would also hold true for the transport and localization of subcellular organelles within the cell.

Using the cellular railroad system, once secretory vesicles dock at the porosome base following a secretory stimulus, the fusion of membrane-bound secretory vesicles at the porosome base is mediated by calcium and a specialized set of three soluble *N*-ethylmaleimide-sensitive factor (*NSF*)-attachment protein receptors called SNAREs [58–62]. In neurons, for example, target membrane proteins SNAP-25 and syntaxin called t-SNAREs present at the base of neuronal porosomes at the presynaptic membrane, and a synaptic vesicle-associated membrane protein (VAMP) or v-SNARE is part of the conserved protein complex involved in membrane fusion and neurotransmission. In the presence of Ca^{2+}, t-SNAREs and v-SNARE in opposing membrane bilayers interact and self-assemble in a ring conformation to form a conducting channel [63]. Such self-assembly of t-/v-SNARE rings occur only when the respective SNAREs are membrane associated [63]. The size of the SNARE ring complex is dependent on the curvature of the opposing lipid membrane [64]. Electron density map and 3D topography of the SNARE ring complex suggests the formation of a leak-proof channel measuring 25 Å in ring thickness, and 42 Å in height [65]. These structural studies combined with functional electrophysiological measurements have greatly advanced our understanding of membrane-directed SNARE ring complex assembly [63–65]. Additionally, the SNARE ring size can now also be mathematically predicted [65]. X-ray diffraction measurements and simulation studies have further advanced that membrane-associated t-SNAREs and v-SNARE overcome repulsive forces to bring the opposing membranes closer to within a distance of approximately 2.8 Å [58, 66, 67]. Calcium is then able to bridge the closely opposed bilayers, leading to the release of water from hydrated Ca^{2+} ions as well as the loosely coordinated water at membrane phospholipid head groups, resulting in membrane destabilization and fusion [67].

Studies demonstrate that during cell secretion, secretory vesicle swelling is required for the expulsion of intravesicular contents [68]. Live pancreatic acinar cells in near physiological buffer when imaged using AFM at high force (200–300 pN) demonstrate the size and shape of secretory vesicles called zymogen granules (ZGs) lying immediately below the apical plasma membrane of the cell. Within 2.5 min of exposure to a secretory stimulus, a majority of ZGs within the acinar cells swell, followed by secretion and the concomitant decrease in ZG size. There is no loss of secretory vesicles during the entire secretory process and following secretion. These studies reveal for the first time in live cells intracellular swelling of secretory vesicles following stimulation of cell secretion and their deflation following partial discharge of vesicular contents [68]. Since no loss of secretory vesicles is observed through the entire secretory process, further demonstrates transient fusion, as opposed to a complete merger of secretory vesicles at the cell plasma membrane. Measurements of intracellular ZG size reveal that different vesicles swell differently following a secretory stimulus. This differential swelling among secretory vesicles within the same cell may explain why following stimulation of cell secretion, some secretory vesicles demonstrate the presence of less vesicular content than others, reflecting variations between them in content discharged. To determine precisely the role of swelling in vesicle–plasma membrane fusion and in intravesicular content expulsion, an electrophysiological ZG-reconstituted lipid bilayer fusion assay has been used [63, 68]. The ZGs used in the bilayer fusion assays are first characterized for their purity and their ability to respond to a swelling stimulus, GTP. As reported [69, 70], exposure of isolated ZGs to GTP results in ZG swelling. Similar to what is observed in live acinar cells, it is demonstrated that every isolated ZG responds differently to the same swelling stimulus. This differential response of isolated ZGs to GTP has been further assessed by measuring percent change in volume of isolated ZGs of different sizes [68]. ZGs in the exocrine pancreas range in size from 0.2 to 1.3 μm in diameter [69], not all ZGs are found to swell following a GTP challenge [68]. Volume increases in most ZGs following GTP exposure is between 5% and 20%; however, larger increases of up to 45% have been reported in vesicles ranging from 250 to 750 nm in diameter. In the electrophysiological bilayer fusion assay, immunoisolated porosome complex from the exocrine pancreas are functionally reconstituted [14] into the lipid membrane of the bilayer apparatus, where membrane conductance and capacitance are continually monitored [68]. Reconstitution of the porosome into the lipid membrane results in a small increase in capacitance, resulting from an increase in membrane surface area. Addition of isolated ZGs to the *cis* compartment of the bilayer chamber results in vesicle docking and fusion at the porosome-reconstituted lipid membrane, detected as a step increase in membrane capacitance. Even after 15 min of ZG addition to the *cis* compartment of the bilayer chamber, little or no release of the intravesicular enzyme α-amylase is detected in the *trans* compartment of the bilayer chamber. By contrast, exposure of ZGs to 20 μM GTP induces swelling and results both in the potentiation of fusion and in a robust expulsion of α-amylase into the *trans* compartment of the bilayer chamber determined using immunoanalysis. These studies demonstrate that during cell secretion, secretory vesicle swelling is required for the precise and regulated expulsion of intravesicular contents. This mechanism of vesicular expulsion during cell secretion may explain how partially empty vesicles are generated in cells following secretion. The presence of empty secretory vesicles could result from multiple rounds of fusion–swelling–expulsion cycles a vesicle may undergo during the secretory process, reflecting on the precise and regulated nature of process.

Discovery of the porosome, its functional reconstitution in artificial lipid membrane, and an understanding of its morphology, composition, and dynamics; the molecular underpinnings of SNARE-induced membrane fusion, and secretory vesicle swelling involved in content expulsion, have resulted in a paradigm shift in our understanding of the secretory process in cells. These new findings in the past 15 years are briefly summarized in this *Springer Brief.*

Materials and Methods

Pancreatic Acini Isolation

Pancreatic acinar cells and hemi-acini were isolated using a published procedure [6]. For each experiment, a male Sprague–Dawley rat weighing 80–100 g was euthanized by ether inhalation. The pancreas was then dissected out and chopped into 0.5-mm^3 pieces, which were then mildly agitated for 15 min at 37°C in a siliconized glass tube with 5 ml of oxygenated buffer A (98 mM NaCl, 4.8 mM KCl, 2 mM CaCl$_2$, 1.2 mM MgCl$_2$, 0.1% bovine serum albumin, 0.01% soybean trypsin inhibitor, 25 mM HEPES, pH 7.4) containing 1,000 units of collagenase. The suspension of acini was filtered through a 224-mm Spectra-Mesh (Spectrum Laboratory Products) polyethylene filter to remove large clumps of acini and undissociated tissue. The acini were washed six times, 5 ml per wash, with ice-cold buffer A. Isolated rat pancreatic acini and acinar cells were plated on Cell-Tak-coated glass coverslips. Two to three hours after plating, cells were imaged using the AFM before and during stimulation of secretion. Isolated pancreatic acinar cells and small acinar preparations were used in the study since fusions of regulated secretory vesicles at the plasma membrane in pancreatic acini are confined to the apical region which faces the acinus lumen, preventing imaging by the AFM in whole acinar preparations. Furthermore, in isolated acinar cells or hemi-acinar preparations, the secretagogue Mas7 has immediate and uniform access to the cells.

Zymogen Granule Isolation

Zymogen granules (ZGs) were isolated according to a minor modification of a published procedure [69]. The pancreas from male Sprague–Dawley rats was dissected and diced into 0.5-mm^3 pieces before being suspended in 15% (w/v) ice-cold homogenization buffer (0.3 M sucrose, 25 mM HEPES, pH 6.5, 1 mM benzamidine, 0.01% soybean trypsin inhibitor) and homogenized using three strokes of a Teflon glass homogenizer. The homogenate was centrifuged for 5 min at 300×g at 4°C. The supernatant fraction was mixed with 2 vol. of a Percoll–sucrose–HEPES buffer (0.3 M sucrose, 25 mM HEPES, pH 6.5, 86% Percoll, 0.01% soybean trypsin inhibitor) and centrifuged for 30 min at 16,400×g at 4°C. Pure ZGs were obtained as a loose white pellet at the bottom of the centrifuge tube and used in the study.

Isolation of Synaptosomes, Synaptosomal Membrane, and Synaptic Vesicles

Rat brain synaptosomes, synaptosomal membrane, and synaptic vesicles were prepared from whole brain tissue of Sprague–Dawley rats. Brains from Sprague–Dawley rats weighing 100–150 g were isolated and placed in an ice-cold buffered sucrose solution (5 mM HEPES pH 7.4, 0.32 M sucrose) supplemented with protease inhibitor cocktail (Sigma, St. Louis, MO, USA) and homogenized using Teflon–glass homogenizer (ten strokes). The total homogenate was centrifuged for 3 min at $2,500 \times g$. The supernatant fraction was further centrifuged for 15 min at $14,500 \times g$, and the resultant pellet was resuspended in buffered sucrose solution, which was loaded onto 3–10–23% Percoll gradients. After centrifugation at $28,000 \times g$ for 6 min, the enriched synaptosomal fraction was collected at the 10–23% Percoll gradient interface. To isolate synaptic vesicles and synaptosomal membrane, isolated synaptosomes were diluted with nine volumes of ice-cold H_2O (hypotonic lysis of synaptosomes to release synaptic vesicles) and immediately homogenized with three strokes in Dounce homogenizer, followed by 30-min incubation on ice. The homogenate was centrifuged for 20 min at $25,500 \times g$, and the resultant pellet (enriched synaptosomal membrane preparation) and supernatant (enriched synaptic vesicles preparation) were used in our studies.

Porosome Isolation

To isolate the neuronal fusion pore or porosome complex, SNAP-25-specific antibody conjugated to protein A-sepharose was used. SNAP-25 is present in neurons. One gram of total rat brain homogenate solubilized in Triton/Lubrol solubilization buffer (0.5% Lubrol; 1 mM benzamidine; 5 mM Mg-ATP; 5 mM EDTA; 0.5% Triton X-100, in PBS) supplemented with protease inhibitor mix (Sigma, St. Louis, MO, USA) was used. SNAP-25 antibody conjugated to the protein A-sepharose was incubated with the solubilized homogenate for 1 h at room temperature followed by washing with wash buffer (500 mM NaCl, 10 mM Tris, 2 mM EDTA, pH 7.5). The immunoprecipitated sample attached to the immunosepharose beads was eluted using low pH (pH 3) buffer to obtain the porosome complex.

Similarly, porosome from the acinar cells of the exocrine pancreas were immunoisolated from plasma membrane preparations, using a SNAP-23-specific antibody (SNAP-23 is present in pancreatic acinar cells). To isolate the porosome complex for reconstitution experiments, SNAP-23-specific antibody conjugated to protein A-sepharose was used. Isolated pancreatic plasma membrane preparations were solubilized in Triton/Lubrol solubilization buffer, supplemented with protease inhibitor mix. SNAP-23 antibody conjugated to the protein A-sepharose was incubated with the solubilized membrane for 1 h at room temperature followed by washing with wash buffer and eluted using low pH buffer.

Atomic Force Microscopy

Cells attached to a Cell-Tak-coated glass coverslip were placed in a thermally controlled fluid chamber that allowed both rapid fluid exchange and the direct visualization of the living cells by an inverted microscope. The newly designed BAFM (Digital Instruments, Santa Barbara, CA, USA) was used in conjunction with an inverted optical microscope (Olympus IX70). Images of the plasma membrane in these cells were obtained by the BAFM, working in the "contact" mode and using a very low vertical imaging force <1 to 3 nN. Silicon and silicon nitride tips were used with spring constants of 0.25 and 0.06 N/m, respectively. To determine the effect of force on the plasma membrane topology, control experiments were performed, where a scanning force of several nanonewtons over a 60-min period demonstrated no significant changes at the plasma membrane.

Measurement of Secretion

Secretions from cells were measured either immunochemically or biochemically. For example, insulin secretion from β-cells or amylase secretion from pancreatic acinar cells were measured using western blot analysis. Amylase secreted from acinar cells was also determined biochemically. Exocytosis from acinar cells was measured by determining the percentage of total cellular amylase release following exposure of cells to a secretagogue (Mas7), or cytochalasin B (actin depolymerizing agent). Amylase, one of the major cargos in ZG of the exocrine pancreas, was measured using the Bernfeld procedure. In a typical amylase assay, rat pancreatic acini dissociated as single cells and groups of two to six cells were used. Fifty to seventy-five cells in 200 μl of total reaction mixture (buffer A) in the presence or absence of Mas7 (secretagogue), Mas 17 (control peptide), or cytochalasin B were incubated at room temperature. Following incubation, the cells were centrifuged at $2,000 \times g$ for 2 min in an Eppendorf microcentrifuge. The supernatant containing the secreted amylase was then assayed. Cells in the remaining 100 μl of incubation mixture were sonicated, and the sonicated mixture was diluted and assayed for amylase. From the above measurements, the total cellular amylase and percent release from cells were calculated. Five microliters of the supernatant or lysed cell fractions was added to 95 μl of ice-cold amylase assay buffer (10 mM NaH_2PO_4, 10 mM Na_2HPO_4, 20 mM NaCl) placed in 12×75-mm glass tubes in an ice bath. The reaction was initiated by the addition of 100 μl of a 10 mg/ml potato starch in amylase assay buffer solution. The mixture was vortexed and incubated for 15 min at 37°C. Following the incubation, the mixture was cooled in an ice bath and 400 μl of a color reagent (44 mM 3,5-dinitrosalicylic acid, 200 mM KOH, and 20 mM sodium potassium tartarate) was added. The mixture in glass tubes was covered and lowered into a boiling water bath for 25 min followed by cooling and the addition of 1.4 ml of distilled water. The mixture was then brought to room temperature and transferred to a plastic cuvette, and absorbance at 530 nm was measured with a spectrophotometer (Beckman DU-64).

Transmission Electron Microscopy

Isolated rat pancreatic acini were fixed in 2.5% buffered paraformaldehyde for 30 min and the pellets embedded in Unicryl resin, followed by sectioning at 40–70 nm. Thin sections were transferred to coated specimen TEM grids, dried in the presence of uranyl acetate and methylcellulose, and examined using a transmission electron microscope. For negative staining electron microscopy, purified protein suspensions in PBS were adsorbed to hydrophilic carbon support films that were mounted onto formvar-coated, metal specimen grids (EMS, Ft. Washington, PA, USA). The adsorbed protein was washed in double-distilled water and negatively stained using 1% aqueous uranyl acetate. After the grids were dried in the presence of the uranyl acetate solution, they were examined by transmission electron microscopy. To prevent bleaching by the electron beam, micrographs were obtained on portions of the grid not previously examined.

Similarly for TEM of neuronal tissue, rat brain was perfused with normal saline solution, followed by phosphate buffer (pH 7.4) containing 2.5% glutaraldehyde. After perfusion, the brain was carefully removed and diced into 1-mm^3 pieces. The pieces of brain tissue were post-fixed in phosphate buffer containing 1.5% osmium tetroxide, dehydrated in graded ethanol and acetone, and embedded in araldite. Tissue blocks were appropriately trimmed and the 40–50-nm sections obtained were stained with lead citrate and examined under a JEOL JEM-100C transmission electron microscope.

Dynamic Light Scattering (DLS) to Determine Vesicle Volume Change

Kinetics of change in ZG and synaptic vesicle volume/size changes were monitored by 90° light scattering with excitation and emission wavelength set at 400 nm in a Hitachi F-2000 spectrophotometer. Synaptic vesicles or ZG suspensions were injected into the thermo-regulated cuvette containing 700 µl of a buffer solution (NaCl 140 mM; KCl 2.5 mM; NaH_2PO_4 0.25 mM; KH_2PO_4 0.25 mM; pH 7.4) at 37°C. For example, light scattering was monitored for 1 min following addition of vesicles to either buffer alone (control) or buffer containing the heterotrimeric G_i-protein stimulant mastoparan (experimental) or its nonstimulatory control peptide, Mast-17 (control), and GTP. Values are expressed as percent increase in light scattered over control values.

Photon Correlation Spectroscopy

The size of neuronal porosome was also determined using photon correlation spectroscopy (PCS). PCS measurements were performed using a Zetasizer Nano ZS (Malvern Instruments, UK). In a typical experiment, the size distribution of isolated porosomes was determined using built-in software provided by Malvern Instruments. Prior to determination of porosome size, calibration of instrument was performed using latex spheres of known size. In PCS, subtle fluctuations in the sample scattering intensity are correlated across microsecond time scales. The correlation function is calculated, from which the diffusion coefficient is determined by the instrument. Using Stokes–Einstein equation, hydrodynamics radius can be calculated from diffusion coefficient. The intensity size distribution, which is obtained as a plot of the relative intensity of light scattered by particles in various size classes, is calculated from correlation function using the built-in software. The particle scattering intensity is proportional to the molecular weight squared. Volume distribution, which assigns more realistic weights to both small and big particles, is calculated from the intensity distribution using Mie theory. The transforms of the PCS intensity distribution to volume distributions are obtained using the provided software by Malvern Instruments.

Porosome Functional Reconstitution Assay

Lipid bilayers were prepared using brain phosphatidylethanolamine (PE) and phosphatidylcholine (PC), and dioleoylphosphatidylcholine (DOPC), and dioleylphosphatidyl-serine (DOPS), obtained from Avanti Lipids (Alabaster, AL, USA). A suspension of PE:PC in a ratio of 7:3 and at a concentration of 10 mg/ml was prepared. Lipid suspension (100 μl) was dried under nitrogen gas and resuspended in 50 μl of decane. To prepare membranes reconstituted with the immunoisolated porosome complex, the immunoisolate was added to the lipid suspension and brushed onto a 200-μm hole in the bilayer preparation cup until a stable bilayer with a capacitance between 100 and 250 pF was established. Alternately, the immunoisolated porosome complex was brushed onto a stable membrane formed at the 200-μm-diameter hole in the bilayer preparation cup. Electrical measurements of the porosome-reconstituted lipid membrane were performed using the EPC9 setup. Current versus time traces were recorded using pulse software, an EPC9 amplifier and probe from HEKA (Lambrecht, Germany). Briefly, membranes were formed while holding at 0 mV. Once a bilayer was formed and demonstrated to be in the capacitance limits for a stable bilayer membrane according to the opening diameter, the voltage was switched to −60 mV. A baseline current was established before the addition of isolated ZGs in case of pancreatic porosomes or synaptic vesicles for reconstituted porosomes from neurons.

Preparation of Liposomes and SNARE Reconstitutions

Purified recombinant SNAREs were reconstituted into lipid vesicles using mild sonication. Three hundred microliters of PC:PS, 100 μl ergosterol, and 15 μl of nystatin (Sigma Chemical Company, St. Louis, MO, USA) were dried under nitrogen gas. The lipids were resuspended in 543 μl of 140 mM NaCl, 10 mM HEPES, and 1 mM CaCl$_2$. The suspension was vortexed for 5 min, sonicated for 30 s, and aliquoted into 100 μl samples (AVs). Twenty-five microliters of syntaxin 1A-1 and SNAP-25 (t-SNAREs) at a concentration of 25 μg/ml was added to 100 μl of AVs. The t-SNARE vesicles were frozen and thawed three times and sonicated for 5 s before use. Bilayer bath solutions contained 140 mM NaCl and 10 mM HEPES. KCl at a concentration of 300 mM was used as a control for testing vesicle fusion.

Circular Dichroism Spectroscopy

Overall secondary structural content of SNAREs and their complexes, both in suspension and in association with membrane, were determined by CD spectroscopy using an Olis DSM 17 spectrometer. Data were acquired at 25°C with a 0.01-cm path length quartz cuvette (Helma). Spectra were collected over a wavelength range of 185–260 nm using 1-nm step spacing. In each experiment, 30 scans were averaged per sample for enhanced signal to noise, and data were acquired on duplicate independent samples to ensure reproducibility. SNAREs and their complexes, both in suspension and in association with membrane, were analyzed for the following samples: v-SNARE, t-SNAREs, v-SNARE + t-SNAREs, v-SNARE + t- SNAREs + N-ethylmaleimide sensitive factor (NSF), and v-SNAR E + t-SNAREs + NSF + 2.5 mM ATP. All samples had final protein concentrations of 10 μM in 5 mM sodium phosphate buffer at pH 7.5 and were baseline subtracted to eliminate buffer (or liposome in buffer) signal. Data were analyzed using the GLOBALWORKS software (Olis), which incorporates a smoothing function and fit using the CONTINLL algorithm.

Wide-Angle X-ray Diffraction

Ten microliters of a 10-mM lipid vesicle suspension was placed at the center of an X-ray polycarbonate film mounted on an aluminum sample holder and placed in a Rigaku RU2000 rotating anode X-ray diffractometer equipped with automatic data collection unit (DATASCAN) and processing software (JADE). Similarly, X-ray diffraction studies were also performed using t- and v-SNARE-reconstituted liposomes, both in the presence and in the absence of Ca^{2+}. Samples were scanned with a rotating anode, using the nickel-filtered Cu Kα line ($\lambda = 1.5418$ Å) operating at 40 kV and 150 mA. Diffraction patterns were recorded digitally with a scan rate of 3°/min using a scintillation counter detector. The scattered X-ray intensities were evaluated as a function of scattering angle 2θ and converted into Å units, using the formula $d(\text{Å}) = \lambda/2\sin\theta$.

Porosome Discovery

The resolving power of the light microscope is dependent on the wavelength of used light and hence 250–300 nm in lateral and much less in depth resolution can at best be achieved using light for imaging. The porosomes in exocrine and neuroendocrine cells are cup-shaped supramolecular structures, measuring 100–180 nm at the opening and 25–45 nm in relative depth. At the nerve terminal or in astrocytes, the porosomes are an order of magnitude smaller, cup-shaped structures measuring just 10–17 nm at the opening to the outside. Due to the nanometer size of the porosome complex, it had evaded visual detection until its discovery using ultrahigh-resolution atomic force microscopy (AFM) [33, 71, 72]. The development of the AFM [71] has enabled the imaging of live cell structure and dynamics in physiological buffer solutions, at nanometer to near angstrom resolution, in real time. In AFM, a probe tip microfabricated from silicon or silicon nitride and mounted on a cantilever spring is used to scan the surface of the sample at a constant force. Either the probe or the sample can be precisely moved in a raster pattern using a *xyz* piezo to scan the surface of the sample. The deflection of the cantilever measured optically is used to generate an isoforce relief of the sample [72]. Force is thus used by the AFM to image surface profiles of objects such as live cells [6–11], subcellular organelles [68–70], and biomolecules [63–65], submerged in physiological buffer solutions, at ultra high resolution and in real time.

Porosomes were first discovered in acinar cells of the exocrine pancreas [6]. Exocrine pancreatic acinar cells are polarized secretory cells possessing an apical and a basolateral end. This well-characterized cell of the exocrine pancreas synthesizes digestive enzymes, which are stored within 0.2–1.2 μm in diameter apically located membranous sacs or secretory vesicles, called zymogen granules (ZGs). Following a secretory stimulus, ZGs dock and fuse with the apical plasma membrane to release their contents to the outside. In contrast to neurons, where secretion of neurotransmitters occurs within millisecond of a secretory stimulus, the pancreatic acinar cells secrete digestive enzymes at the apical plasma membrane over several minutes following stimulation and therefore were chosen to determine the molecular steps involved in cell secretion. In the mid-1990s, AFM studies were undertaken on live pancreatic acinar cells to evaluate at high resolution the structure and dynamics of the apical region of the plasma membrane in both resting and stimulated cells. Isolated live pancreatic acinar cells in physiological buffer, when imaged using the AFM, reveal new cellular structures at the apical pole of the cell where secretion is known to occur. A group of circular "pits" measuring 0.4–1.2 μm in diameter which contain smaller 100–180 nm in diameter "depressions" were identified (Fig. 1a–d) at the apical plasma membrane. These "depression structures" were subsequently named "porosome" or the secretory portal in cells. Typically three to four depressions are found within each pit structure at that apical plasma membrane. Not surprisingly, the basolateral membrane of pancreatic acinar cells is devoid of such pit and depression structures [6]. High-resolution AFM images of depressions in live acinar cells further reveal a cone-shaped basket-like morphology, each cone

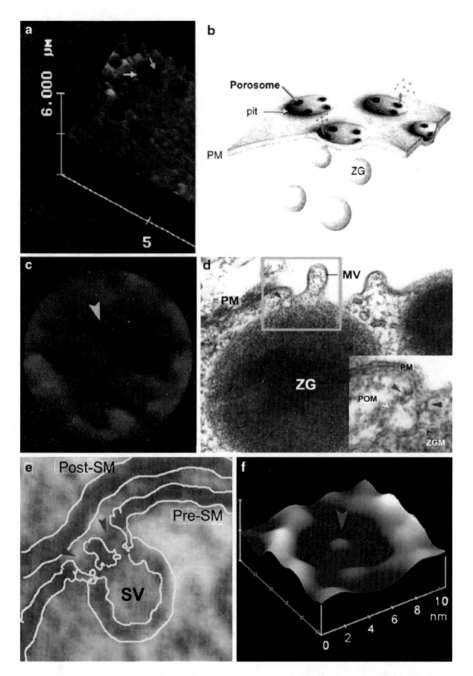

Fig. 1 Porosomes, previously referred to as "depressions" at the plasma membrane in pancreatic acinar cells and at the nerve terminal. (**a**) AFM micrograph depicting "pits" and "porosomes" within at the apical plasma membrane in a live pancreatic acinar cell. (**b**) To the right is a schematic drawing depicting porosomes at the cell plasma membrane (PM), where membrane-bound secretory vesicles called zymogen granules (ZGs) dock and fuse to release intravesicular contents. (**c**) A high-resolution AFM micrograph showing a single pit with four 100–180-nm porosomes within. (**d**) An electron micrograph depicting a porosome (*red arrowhead*) close to microvilli (MV) at the apical

measuring 15–35 nm in depth. Subsequent studies over the years demonstrate the presence of depressions in all secretory cells examined, including neurons (Figs. 1e, f and 2). Analogous to pancreatic acinar cells, examination of resting GH secreting cells of the pituitary [9] and chromaffin cells of the adrenal medulla [8] also reveals the presence of pits and depressions at the cell plasma membrane. The presence of depressions or porosomes in neurons, astrocytes, β-cells of the endocrine pancreas and in mast cells has also been determined, demonstrating their universal presence in secretory cells [18, 19, 24].

Exposure of pancreatic acinar cells to a secretagogue results in a time-dependent increase (20–45%) in both the diameter and relative depth of depressions (Fig. 3). Studies demonstrate that depressions return to resting size on completion of cell secretion [6, 7]. However, no demonstrable change in pit size is detected following stimulation of secretion [6]. Enlargement of depression diameter and an increase in its relative depth after exposure to secretagogue correlate with secretion. Additionally, exposure of pancreatic acinar cells to cytochalasin B, a fungal toxin that inhibits actin polymerization and secretion, results in a 15–20% decrease in depression size and a consequent 50–60% loss in secretion [6]. Results from these studies were the first to suggest depressions to be the secretory portals in pancreatic acinar cells. Furthermore, these studies demonstrated the involvement of actin in regulation of both the structure and function of depressions. Similar to pancreatic acinar cells, depression in resting GH cells measure 154 ± 4.5 nm (mean $\pm$ SE) in diameter, and following exposure to a secretagogue results in a 40% increase in depression diameter (215 ± 4.6 nm; $p < 0.01$), with no appreciable change in pit size [9]. The enlargement of depression diameter during cell secretion and its subsequent decrease, accompanied by the loss in secretion following exposure to actin depolymerizing agents [9], also suggested them to be the secretory portal in GH cells. A direct determination that depressions are the secretory portals in cells, via which secretory products are expelled, was unequivocally demonstrated using immuno-AFM studies first in the exocrine pancreas [7] (Fig. 4), followed by studies in the GH cells of pituitary [9]. Localization at depressions, gold-conjugated antibody to secretory proteins, finally provided direct evidence that secretion occur through depressions. ZGs contain the starch-digesting enzyme amylase. AFM micrographs of the specific

---

Fig. 1 (continued) plasma membrane (PM) of a pancreatic acinar cell. Note the association of the porosome membrane (*yellow arrowhead*), and the zymogen granule membrane (ZGM) (*red arrowhead*) of a docked ZG (*inset*). Cross section of a circular complex at the mouth of the porosome is seen. (**e**) The *bottom left panel* shows an electron micrograph of a porosome at the nerve terminal, in association with a synaptic vesicle (SV) at the presynaptic membrane (Pre-SM). Notice a central plug at the neuronal porosome opening. (**f**) The *bottom right panel* is an AFM micrograph of a neuronal porosome in physiological buffer, also showing the central plug at its opening. It is believed that the central plug in neuronal porosomes may regulate its rapid close–open conformation during neurotransmitter release. The neuronal porosome is an order of magnitude smaller (10–15 nm) in comparison with porosomes in the exocrine pancreas. (Figure represents a collage of images from our earlier publications: *Proc Natl Acad Sci* 1997, 94:316–321; *Biophys J* 2003, 85:2035–2043; *Cell Biol Int* 2004, 28:699–708). ©Bhanu Jena

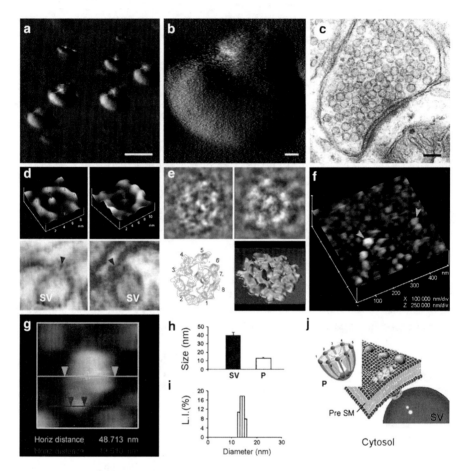

Fig. 2 Structure and organization of the neuronal porosome complex at the nerve terminal. (a) Low-resolution AFM amplitude image (bar = 1 μm) and (b) high-resolution AFM amplitude image (bar = 100 nm) of isolated rat brain synaptosomes in buffered solution. (c) Electron micrograph of a synaptosome (bar = 100 nm). (d) Structure and arrangement of the neuronal porosome complex facing the outside (*top left*), and the arrangement of the reconstituted complex in PC:PS membrane (*top right*). *Lower panels* depict two transmission electron micrographs demonstrating synaptic vesicles (SV) docked at the base of a cup-shaped porosome, having a central plug (*red arrowhead*). (e) EM, electron density, and 3D contour mapping demonstrate at the nanoscale the structure and assembly of proteins within the complex. (f) AFM micrograph of inside-out membrane preparations of isolated synaptosome. Note the porosomes (*red arrowheads*) to which synaptic vesicles are found docked (*blue arrowhead*). (g) High-resolution AFM micrograph of a synaptic vesicle docked to a porosome at the cytoplasmic compartment of the presynaptic membrane. (h) AFM measurements (n = 15) of porosomes (P, 13.05 ± 0.91) and synaptic vesicles (SV, 40.15 ± 3.14) at the cytoplasmic compartment of the presynaptic membrane. (i) Photon correlation spectroscopy (PCS) of immunoisolated neuronal porosome complex demonstrating a size of 12–16 nm. (j) Schematic illustration of a neuronal porosome at the presynaptic membrane, showing the eight ridges connected to the central plug. (Figure represents a collage of images from our previous publications: *Cell Biol Int* 2004, 28:699–708; *Cell Biol Int* 2010, 34:1129–1132; *J Microsc.* 2008, 232:106–111). ©Bhanu Jena

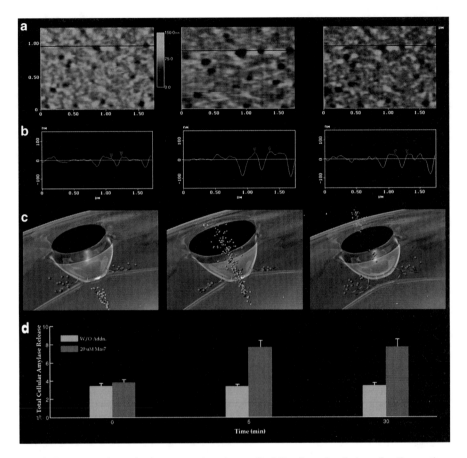

Fig. 3 Porosome dynamics in pancreatic acinar cells following stimulation of cell secretion. (**a**) Several porosomes within a pit are shown at zero time, 5, and 30 min following stimulation of secretion. (**b**) Section analysis across three porosomes in the *top panel* is represented graphically in the *second panel* and defines the diameter and relative depth of each of the three porosomes. The porosome at the center is represented by *red arrowheads*. (**c**) The *third panel* is a 3D rendition of the porosome complex at different times following stimulation of secretion. Note the porosome as a *blue cup*-shaped structure with *black opening* to the outside, and part of a secretory vesicle (*violet*) docked at its base via t-/v-SNAREs. (**d**) The *bottom panel* represents % total cellular amylase release in the presence and absence of the secretagogue Mas7 (*blue bars*). Note an increase in porosome diameter and relative depth, correlating with an increase in total cellular amylase release at 5 min following stimulation of secretion. At 30 min following a secretory stimulus, there is a decrease in diameter and relative depth of porosomes and no further increase in amylase release beyond the 5-min time point. No significant changes in amylase secretion (*green bars*) or porosome diameter were observed in control cells in either the presence or absence of the nonstimulatory mastoparan analog (Mas17). High-resolution images of porosomes were obtained before and after stimulation with Mas7, for up to 30 min. [Modified figure from our earlier publication: *Proc Natl Acad Sci* 1997, 94:316–321]. ©Bhanu Jena

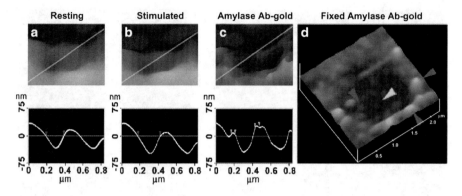

Fig. 4 Intravesicular contents are expelled to the outside through the porosome during cell secretion. (**a, b**) AFM micrograph and section analysis of a pit and two of the four depressions or porosomes, demonstrating enlargement of porosomes following stimulation of cell secretion in the acinar cell of the exocrine pancreas. (**c**) Exposure of live cells to gold-conjugated amylase antibody (Ab) results in specific localization of gold particles to these secretory sites. Note the localization of amylase-specific immunogold particles at the edge of porosomes. (**d**) AFM micrograph of pits and porosomes with immunogold localization demonstrated in cells immunolabeled and then fixed. *Blue arrowheads* point to immunogold clusters and the *yellow arrowhead* points to a depression or porosome opening. (Figure from our earlier publication: *Cell Biol Int* 2002, 26:35–42). ©Bhanu Jena

localization of gold-tagged amylase-specific antibodies at depressions following stimulation of cell secretion [7, 13] conclusively demonstrated depressions to be the secretory portal in cells. Similarly in somatotrophs of the pituitary gland, gold-tagged growth hormone-specific antibody found to selectively localize at the depression openings following stimulation of secretion [9], which established these sites to represent the secretory portals in GH cells. Porosomes have also been identified and dynamics of their structure and function examined in insulin-secreting β-cells of the endocrine pancreas (Fig. 5, Jena, personal observation). In pancreatic β-cells, porosomes range in size from 100 to 130 nm. Similar to GH and pancreatic acinar cells, exposure of β-cells to a secretagogue (25 mM glucose) results in dilation of the porosome and insulin secretion (Fig. 5). Exposure of β-cells to 20 μM cytochalasin B results in the decrease in porosome opening and a concomitant loss in insulin secretion (Fig. 5). Over the years, the term "fusion pore" has been loosely referred to plasma membrane dimples that originate following a secretory stimulus, or to the continuity or channel established between opposing lipid membrane during membrane fusion. Therefore, for clarity, the term "porosome" was assigned to these depression structures at the cell plasma membrane where secretion occurs.

The porosome structure, at the cytosolic compartment of the plasma membrane in the exocrine pancreas ([15], Fig. 6), and in neurons ([10], Fig. 7), has also been determined at near nanometer resolution in live cells. To determine the morphology of porosomes at the cytosolic compartment of pancreatic acinar cells, isolated plasma membrane preparations in near physiological buffered solution have been imaged at ultrahigh resolution using the AFM. These studies reveal scattered

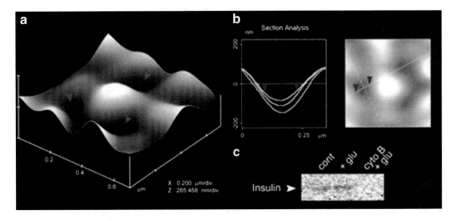

Fig. 5 Porosome dynamics in live rat pancreatic β cells. (**a**) AFM micrographs of porosomes (*red arrowheads*) measuring 100–130 nm, at the surface of the apical plasma membrane in a live rat pancreatic β cell. (**b**) Section analysis of a porosome (*right red arrowhead*) in a resting live rat pancreatic β cell (*yellow line*), dilated (*green line*) following stimulation of secretion using 25 mM glucose, and decrease in porosome size (*white line*) following exposure to the actin cytochalasin B. (**c**) Insulin-immunoblot analysis of medium containing live rat pancreatic β cells in low glucose (5 mM or control), stimulated using 25 mM glucose, and cells exposed to cytochalasin B prior to glucose stimulation. Note the loss of 25 mM glucose-stimulated insulin secretion in cells pre-exposed to cytochalasin B. ©Bhanu Jena

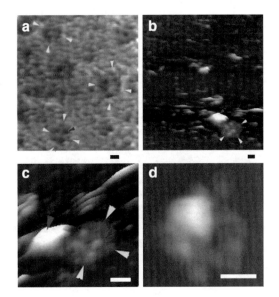

Fig. 6 AFM micrographs of porosomes within "pits" at the surface of the apical plasma membrane in live pancreatic acinar cells, and at the cytosolic compartment of an isolated pancreatic plasma membrane preparation. (**a**) Several circular "pits" (*yellow arrowheads*) with porosomes within (*red arrowheads*) are seen in this AFM micrograph of the apical plasma membrane in a live pancreatic acinar cell. (**b**) AFM micrograph of the cytosolic compartment of an isolated pancreatic plasma membrane preparation depicting a "pit" (*yellow arrowheads*) containing several cup-shaped porosome structures (*red arrowhead*) within, associated with a ZG (*blue arrowhead*). (**c**) The "pit" and inverted porosomes in B are shown at higher magnification. (**d**) AFM micrograph of another "pit" with inverted porosomes within and associated ZG (bar = 200 nm). (Figure from our earlier publication: *Biophys J* 2003, 85:2035–2043). ©Bhanu Jena

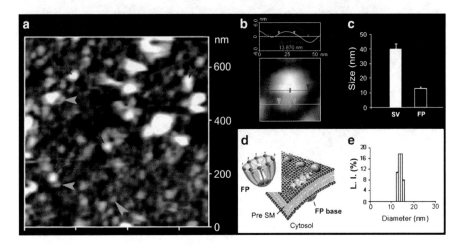

Fig. 7 Neuronal fusion pore distribution, size, and structure. (**a**) Figure shows the structure and distribution of fusion pores at the cytosolic compartment of a synaptosome. Inside-out synaptosome preparations when imaged in buffer using AFM demonstrate inverted 12–16 nm cup-shaped fusion pores, some with docked vesicles. Note one inverted cup-shaped fusion pore (*green arrowheads*), with a docked synaptic vesicle (*red arrowheads*), shown at higher magnification in "**b**." (**b**) AFM micrograph shows a 37-nm synaptic vesicle docked to a 14-nm fusion pore at the cytoplasmic compartment in the isolated synaptosomal membrane. (**c**) AFM measurement of the fusion pores (13.05 ± 0.91) and attached synaptic vesicles (40.15 ± 3.14) in the cytosolic compartment of the synaptosome membrane (*n* = 15). (**d**) Schematic illustration of a neuronal fusion pore showing the eight vertical ridges and a central plug. (**e**) Photon correlation spectroscopy further demonstrates fusion pores measuring 12–16 nm. (Figure from our earlier publication: *Cell Biol. Int.* 2007, 31:1301–1308). ©Bhanu Jena

circular disks measuring 0.5–1 μm in diameter, with inverted cup-shaped structures within [14]. The inverted cups at the cytosolic compartment of isolated pancreatic plasma membrane preparations measure approximately 15 nm in height. On a number of occasions, ZGs ranging in size from 0.4 to 1 μm in diameter are observed in association with one or more of the inverted cups, suggesting the circular disks to represent pits, and inverted cups porosomes, in inside-out pancreatic plasma membrane preparations. To further confirm that the cup-shaped structures are porosomes where secretory vesicles dock and fuse, immuno-AFM studies have been carried out. Target membrane proteins SNAP-23 [61, 73] and syntaxin [62] (t-SNARE), and secretory vesicle-associated membrane protein v-SNARE or VAMP [60], are part of the conserved protein complex involved in fusion of opposing bilayers in cells [58, 66, 67]. Since ZGs dock and fuse at the plasma membrane to release vesicular contents, it was hypothesized that if the inverted cups or porosomes are the secretory portals, then plasma membrane-associated t-SNAREs should localize at the base of the cup-shaped structure. The t-SNARE protein SNAP-23 had previously been reported in pancreatic acinar cells [73]. A polyclonal monospecific SNAP-23 antibody recognizing a single 23-kDa protein in western blots of pancreatic plasma membrane fraction, when used in immuno-AFM studies, demonstrated selective localization to the base of the cup-shaped structures ([14], Fig. 8).

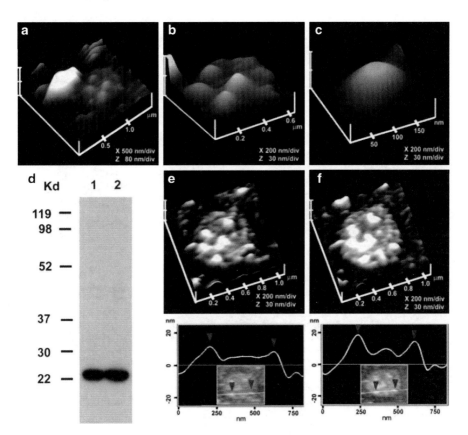

Fig. 8 Morphology of the cytosolic compartment of the porosome complex revealed in AFM studies on isolated pancreatic plasma membrane preparations. (**a**) This AFM micrograph of isolated plasma membrane preparation reveals the cytosolic compartment of a pit with inverted cup-shaped porosomes. Note the 600 nm in diameter ZG at the *left-hand corner* of the pit. (**b**) Higher magnification of the same pit demonstrates the presence of four to five porosomes within. (**c**) The cytosolic side of a single porosome is depicted in this AFM micrograph. (**d**) Immunoblot analysis of 10 and 20 μg of pancreatic plasma membrane preparations, using SNAP-23 antibody, demonstrates a single 23-kDa immunoreactive band. (**e, f**) The cytosolic side of the plasma membrane demonstrates the presence of a pit with a number of porosomes within, shown before (**e**) and after (**f**) addition of the SNAP-23 antibody. Note the increase in height of the porosome base revealed by section analysis (*bottom panel*), demonstrating localization of SNAP-23 antibody to the base of the porosome. (Figure from our earlier publication: *Biophys J* 2003, 84:1–7). ©Bhanu Jena

These results confirm that the inverted cup-shaped structures in inside-out pancreatic plasma membrane preparations are indeed porosomes where secretory vesicles transiently dock and fuse to release their contents during cell secretion. The size, shape, and 3D contour map of immunoisolated porosome complex has also been determined using both negative staining EM and AFM studies in exocrine pancreas ([14], Fig. 9), neurons ([10–12], Fig. 2), and astrocytes [74]. The immunoisolated porosome complex has further been structurally and functionally reconstituted into

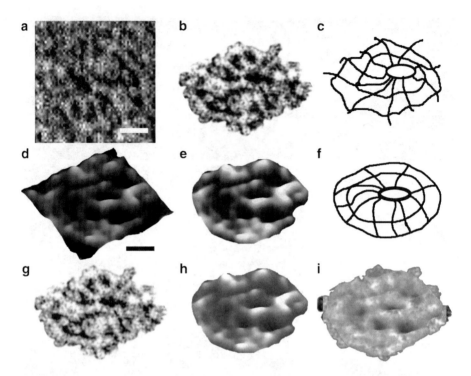

Fig. 9 Electron micrograph from negatively stained preparation and AFM micrographs of immunoisolated porosome complex from the exocrine pancreas. (**a**) Negatively stained electron micrograph of an immunoisolated porosome from solubilized pancreatic plasma membrane preparations, using a SNAP-23-specific antibody. Note the three rings and the ten spokes that originate from the innermost small ring. This structure represents the protein backbone of the porosome complex. The three rings and the vertical spikes are also observed in electron micrographs of intact cells and in porosomes co-isolated with ZGs (bar = 30 nm). (**b**) Electron micrograph of the porosome complex depicted in "**a**," and (**c**) an outline of the structure presented for clarity. (**d–f**) AFM micrograph of isolated porosome complex in near physiological buffer (bar = 30 nm). Note the structural similarity of the complex, imaged both by TEM (**g**) and AFM (**h**). The TEM and AFM micrographs are superimposable (**i**). (Figure from our earlier publication: *Biophys J* 2003, 85:2035–2043). ©Bhanu Jena

artificial liposomes and lipid bilayer membrane ([10–12, 14], Fig. 10). Transmission electron micrographs of pancreatic porosomes reconstituted into liposomes exhibit a 150–200-nm cup-shaped basket-like morphology, similar to its native structure observed in cells and when co-isolated with ZG preparations [14]. To test the functionality of isolated porosomes, purified porosome preparations obtained from exocrine pancreas or neurons have been reconstituted in lipid membrane of the electrophysiological bilayer setup (EPC9) and exposed to isolated ZGs (Fig. 10) or synaptic vesicle preparations. Electrical activity of the porosome-reconstituted membrane, as well as the transport of vesicular contents from the *cis* to the *trans* compartments of the bilayer chambers when monitored, demonstrates that the lipid membrane-reconstituted porosomes are indeed functional [10, 14], since in the presence of calcium, isolated secretory vesicles dock and transiently fuse at the

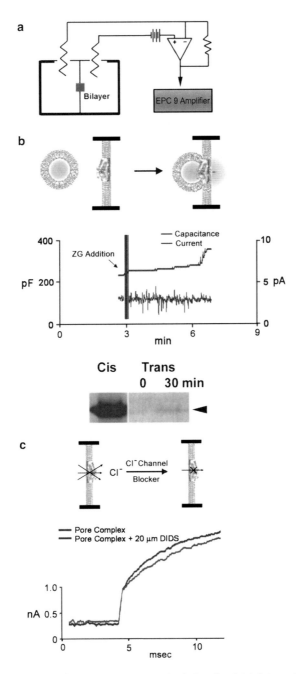

Fig. 10 Lipid bilayer-reconstituted porosome complex is functional. (**a**) Schematic drawing of the bilayer setup for electrophysiological measurements. (**b**) Zymogen granules (ZGs) added to the *cis* compartment of the bilayer fuse with the reconstituted porosomes, as demonstrated by an increase in capacitance and current activities, and a concomitant time-dependent release of amylase (a major ZG content) to the *trans* compartment of the membrane. The movement of amylase from the *cis* to the *trans* compartment of the chamber was determined by immunoblot analysis of the contents in the *cis* and the *trans* chamber over time. (**c**) As demonstrated by immunoblot analysis of the immunoisolated complex, electrical measurements in the presence and absence of the chloride ion channel blocker DIDS indicate the presence of chloride channels in association with the complex. (Figure from our earlier publication: *Biophys J* 2003, 85:2035–2043). ©Bhanu Jena

porosome to transfer intravesicular contents from the *cis* to the *trans* compartment of the bilayer chamber. A time-dependent transport of the ZG enzyme amylase from *cis* to the *trans* compartment of the bilayer chamber is demonstrated using immuno-blot analysis of the buffer in the *cis* and *trans* compartments of the bilayer chambers. In pancreas, chloride channel activity present at the reconstituted porosome complex is critical to its activity, since the chloride channel blocker DIDS was found to inhibit function of the reconstituted porosome (Fig. 10).

Similarly, the structure and biochemical composition of the neuronal porosome and the docking and fusion of synaptic vesicles at the neuronal porosome complex have been demonstrated. AFM, EM, and electron density measurements followed by contour mapping, and 3D topography of the neuronal porosome, further provided an understanding of the arrangement of proteins at nanometer resolution within the complex ([12], Fig. 10). Results from these studies demonstrate that proteins at the central plug of the porosome interact with proteins at the periphery of the complex, conforming to its eightfold symmetry (Fig. 2d, e). Furthermore, at the center of the porosome complex representing the porosome base, where synaptic vesicles dock and transiently fuse, SNARE proteins are assembled in a ring conformation. In neurons, this SNARE ring at the porosome base is composed of merely three SNARE pairs [75, 76] having a 1–1.5 nm in diameter channel, for the express release of neurotransmitters from fused synaptic vesicles via the porosome to the synaptic cleft. These studies demonstrate that porosomes are permanent structures at the presynaptic membrane of nerve terminals, where synaptic vesicles transiently dock and fuse to release neurotransmitters. Photon correlation spectroscopy (PCS) of isolated porosome complexes further confirms that neuronal porosomes measure on average 12–15 nm (Fig. 2i). In PCS measurements, the size distribution of isolated porosome complexes is obtained from plots of the relative intensity of light scattered by particles of known sizes and a calculation of their correlation function. Negative staining EM performed using low electron dose, in a Tacnai 20 electron microscope operating at 200 kV, further confirms the porosome size and demonstrates that proteins at the central plug of the porosome complex interact with proteins at the periphery of the structure [12]. Similar to AFM micrographs, approximately eight interconnected protein densities are observed at the lip of the porosome complex in EM micrographs (Fig. 2). The eight interconnected protein densities are also connected to the central plug, via spoke-like elements. Electron density and contour maps, and the resultant 3D topology profiles of the porosome complex, provide further details of the circular arrangement of proteins, and their connection to the central plug via distinct spoke-like structures (Fig. 2e). The contour map of proteins within the neuronal porosome complex has been determined using published approaches and procedures [77–80]. Results from these studies provide the arrangement of proteins at the nanometer scale within the neuronal porosome complex. The next level of understanding of this supramolecular structure requires electron crystallography on the isolated complex, which are currently under way.

Biochemical analysis of the isolated porosome demonstrates the complex to be composed of SNAP, syntaxin, cytoskeletal proteins actin, α-fodrin, and vimentin, calcium channels β3 and α1c, together with the SNARE regulatory protein NSF [13, 14].

Chloride ion channels ClC2 and ClC3 have also been identified as part of the porosome complex, critical to its function. Isoforms of the various proteins identified within the porosome complex have also been determined using 2D-BAC gels electrophoresis. For example, three isoforms each of the calcium ion channel and vimentin are found in porosomes. Using yeast two-hybrid analysis and immunoisolation studies, the presence and direct interaction between some of these proteins with t-SNAREs within the porosome complex have been established [81]. Besides proteins, studies report that the neuronal porosome assembly requires membrane cholesterol [11]. Results from studies [11] demonstrate a significant inhibition in interactions between porosome-associated t-SNAREs and calcium channels following depletion of membrane cholesterol. Since calcium is critical to SNARE-induced membrane fusion, the loss of interaction between SNAP-25, Syntaxin-1, and calcium channel at the neuronal porosome complex would seriously compromise or even abrogate neurotransmission at the nerve terminal.

In summary, these studies demonstrate porosomes to be permanent supramolecular lipoprotein structures at the cell plasma membrane where membrane-bound secretory vesicles transiently dock and fuse to release intravesicular contents to the outside. Porosomes, therefore, are the universal secretory portals in cells [15–42].

Calcium and SNARE-Induced Membrane Fusion

In the past two decades, much progress has been made in our understanding of membrane fusion in cells, beginning with the discovery of an N-ethylmaleimide-sensitive factor (NSF) [59] and SNARE proteins [60–62], and the determination of their participation in membrane fusion [58, 63, 64, 66, 82–84]. VAMP and syntaxin are both integral membrane proteins, with the soluble SNAP-25 associating with syntaxin. Therefore, the understanding of SNARE-induced membrane fusion requires determining the atomic arrangement and interactions between membrane-associated v- and t-SNARE proteins. Ideally, the atomic coordinates of membrane-associated SNARE complex using X-ray crystallography would help to elucidate the chemistry of SNARE-induced membrane fusion in cells. So far such structural details at the atomic level of membrane-associated t-/v-SNARE complex have not been possible, primarily due to solubility problems of membrane-associated SNAREs, compounded with the fact that v-SNARE and t-SNAREs need to reside in opposing membranes when they meet, to assemble in a physiologically relevant SNARE complex. The remaining option is the use of nuclear magnetic resonance spectroscopy (NMR), which has been of little help, since the size of t-/v-SNARE ring complex is beyond the optimal limit for NMR studies. Regardless, high-resolution AFM force spectroscopy, and EM electron density map and 3D topography of the SNARE ring complex have enabled an understanding of the structure, assembly, and disassembly of membrane-associated t-/v-SNARE complexes in physiological buffer solution [58, 63–66, 82, 83].

The structure and arrangement of SNAREs associated with lipid bilayer were first determined using AFM almost a decade ago [63]. Electrophysiological measurements of membrane conductance and capacitance enabled the determination of fusion of v-SNARE-reconstituted liposomes with t-SNARE-reconstituted membrane. Results from these studies demonstrated that t-SNAREs and v-SNARE when present in opposing membrane interact and assemble in a circular array, and in the presence of calcium, form conducting channels [63]. The interaction of t-/v-SNARE proteins to form such conducting channels is strictly dependent on the presence of t-SNAREs and v-SNARE in opposing membranes. Simple addition of purified recombinant v-SNARE to a t-SNARE-reconstituted lipid membrane fails to form the SNARE ring complex and is without influence on the electrical properties of the membrane [63], i.e., fails to form conducting channels. However, when v-SNARE is reconstituted into liposomes, and these v-SNARE vesicles are added to t-SNARE-reconstituted membrane, SNAREs assemble in a ring conformation, and in the presence of calcium, establish continuity between the opposing membrane bilayers. The establishment of continuity between the opposing t-SNARE- and v-SNARE-reconstituted bilayers is reflected in the increase in membrane capacitance and conductance. These results confirm that t- and v-SNAREs are required to reside in opposing membrane, similar to their presence and function in cells, to allow appropriate t-/v-SNARE interactions leading to the establishment of continuity between opposing membranes [63, 64].

Calcium is critical to SNARE-induced membrane fusion. Studies using SNARE-reconstituted liposomes and bilayers [58] further demonstrate a low fusion rate ($\tau = 16$ min) between t-SNARE-reconstituted and v-SNARE-reconstituted liposomes in the absence of Ca^{2+}. Exposure of t-/v-SNARE liposomes to Ca^{2+} drives vesicle fusion on a near physiological relevant time scale ($\tau \sim 10$ s), demonstrating Ca^{2+} and SNAREs in combination to be the universal fusion machinery in cells [58]. Native and synthetic vesicles exhibit a significant negative surface charge primarily due to the polar phosphate head groups, generating a repulsive force that prevents the aggregation and fusion of opposing vesicles. In cells, SNAREs provide direction and specificity and bring opposing bilayers closer to within a distance of 2–3 Å [58], enabling Ca^{2+} bridging and membrane fusion. The bound Ca^{2+} then leads to the expulsion of water between the bilayers at the bridging site, leading to lipid mixing and membrane fusion. Hence, SNAREs, besides bringing opposing bilayers closer, dictate the site and size of the fusion area during cell secretion. The size of the t-/v-SNARE complex is dictated by the curvature of the opposing membranes [64], and hence smaller the vesicle, the smaller the t-/v-SNARE ring channel formed.

A unique set of chemical and physical properties of the Ca^{2+} ion make it ideal for participating in the membrane fusion reaction. Calcium ion exists in its hydrated state within cells. The properties of hydrated calcium have been extensively studied using X-ray diffraction, neutron scattering, in combination with molecular dynamics simulations [82, 85–87]. The molecular dynamic simulations include three-body corrections compared with ab initio quantum mechanics/molecular mechanics molecular dynamics simulations. First principle molecular dynamics has also been used to investigate the structural, vibrational, and energetic properties of $[Ca(H_2O)_n]^{2+}$ clusters, and the hydration shell of the calcium ion. These studies demonstrate that hydrated calcium $[Ca(H_2O)_n]^{2+}$ has more than one shell around the Ca^{2+}, with the first hydration shell having six water molecules in an octahedral arrangement [85]. In studies using light scattering and X-ray diffraction of SNARE-reconstituted liposomes, it has been demonstrated that fusion proceeds only when Ca^{2+} ions are available between the t- and v-SNARE-apposed proteoliposomes [58, 66]. Mixing of t- and v-SNARE proteoliposomes in the absence of Ca^{2+} leads to a diffuse and asymmetric diffractogram in X-ray diffraction studies, a typical characteristic of short range ordering in a liquid system [86]. By contrast, when t-SNARE and v-SNARE proteoliposomes in the presence of Ca^{2+} are mixed, it leads to a more structured diffractogram, with approximately a 12% increase in X-ray scattering intensity, suggesting an increase in the number of contacts between opposing bilayers, established presumably through calcium-phosphate bridges, as previously suggested [58, 66, 87]. The ordering effect of Ca^{2+} on inter-bilayer contacts observed in X-ray studies [58] is in good agreement with light, AFM, and spectroscopic studies, suggesting close apposition of PO-lipid head groups in the presence of Ca^{2+}, followed by formation of Ca^{2+}–PO bridges between the adjacent bilayers [58, 66, 88]. X-ray diffraction studies show that the effect of Ca^{2+} on bilayer orientation and inter-bilayer contacts is most prominent in the area of 3 Å, with additional appearance of a new peak at position 2.8 Å, both of which are within the ionic radius of Ca^{2+} [58]. These studies further suggest that the ionic radius of Ca^{2+} may make it an ideal player in the membrane fusion reaction. Hydrated calcium $[Ca(H_2O)_n]^{2+}$, however, with a hydration shell having six water molecules and measuring ~6 Å, would be excluded from the t-/v-SNARE-apposed inter-bilayer space; hence, calcium has to be present in the buffer solution when t-SNARE vesicles and v-SNARE vesicles meet. Indeed, studies demonstrate that if t- and v-SNARE vesicles are allowed to mix in a calcium-free buffer, there is no fusion following post-addition of calcium [66]. How does calcium work? Calcium bridging of apposing bilayers may lead to the release of water from the hydrated Ca^{2+} ion, leading to bilayer destabilization and membrane fusion. Additionally, the binding of calcium to the phosphate head groups of the apposing bilayers may also displace the loosely coordinated water at the PO-lipid head groups, resulting in further dehydration, leading to destabilization of the lipid bilayer and membrane fusion. Recent studies in the laboratory [67], using molecular dynamics simulations in the isobaric–isothermal ensemble to determine whether Ca^{2+} was capable of bridging opposing phospholipid head groups in the early stages of the membrane fusion process, demonstrate indeed this to be the case. Furthermore, the distance between the oxygen atoms of the opposing PO-lipid head groups bridged by calcium is in agreement with the 2.8 Å distance previously determined using X-ray diffraction measurements. The hypothesis that there is loss of coordinated water both from the hydrated calcium ion and from the oxygen of the phospholipid head groups in opposing bilayers, following calcium bridging, is further demonstrated from the study.

In the presence of ATP, the highly stable, membrane-directed, and self-assembled t-/v-SNARE complex can be disassembled by a soluble ATPase, the N-ethylmaleimide-sensitive factor (NSF) [82, 83, 89]. Careful examination of the partially disassembled t-/v-SNARE bundles within the complex using AFM demonstrates a left-handed super coiling of SNAREs. These results demonstrate that t-/v-SNARE disassembly

Table 1 Secondary structural fit parameters of SNARE complex formation and dissociation [89]

Protein[a]	Suspension (100×f[b])					Membrane associated (100×f)				
	α	β	O	U	Fit[c]	α	β	O	U	Fit
v-SNARE	4	36	18	43	0.19	0	30	32	38	0.21
t-SNAREs	66	34	0	0	0.02	20	15	21	44	0.84
v-/t-SNAREs	48	52	0	0	0.02	20	19	56	5	0.38
v-/t-SNAREs + NSF	20	25	0	55	0.07	18	6	8	68	0.2
v-/t-SNAREs + NSF + ATP	3	39	18	40	0.22	1	27	34	38	0.23

[a]Protein constructs: v-SNARE, VAMP2; t-SNAREs, SNAP-25 + syntaxin 1A; NSF, N-ethylmaleimide sensitive factor. ATP, adenosine triphosphate
[b]f fraction of residues is a given conformational class; α, α-helix; β, β-sheet; O, other (sum of turns, distorted helix, distorted sheet); U, unordered
[c]Fit, goodness of fit parameter expressed as normalized spectral fit standard deviation (nm)

requires the right-handed uncoiling of each SNARE bundle within the ring complex, demonstrating NSF to behave as a right-handed molecular motor [82]. Furthermore, studies using circular dichroism (CD) spectroscopy [89] report for the first time that both t-SNAREs and v-SNARE and their complexes in buffered suspension exhibit defined peaks at CD signals of 208 and 222 nm wavelengths, consistent with a higher degree of helical secondary structure. Surprisingly, when incorporated in lipid membrane, both SNAREs and their complexes exhibit reduced folding [89]. In conformation with AFM studies, NSF in the presence of ATP disassembles the SNARE complex as reflected from the CD signals, demonstrating elimination of α-helices within the structure ([89], Table 1). These results further demonstrate that NSF–ATP is sufficient for the disassembly of the t-/v-SNARE complex. These studies provide a molecular understanding of SNARE-induced membrane fusion in cells.

v- and t-SNAREs Need to Reside in Opposing Membrane for Appropriate Interactions and the Establishment of Continuity Between Opposing Membranes

Purified recombinant t- and v-SNARE proteins, when applied to a lipid membrane, form globular complexes (Fig. 11a–d) ranging in size from 30 to 100 nm in diameter and 3 to 15 nm in height when examined using AFM. Section analysis of

Fig. 11 (continued) Three-dimensional AFM micrographs of neuronal t-SNAREs deposited on the lipid membrane (**b**), and following the addition of v-SNARE (**c**). Section analysis of the SNARE complex in (**b**) and (**c**) is depicted in (**d**). Note the smaller curve belonging to the t-SNARE complex in (**b**) is markedly enlarged following addition of v-SNARE. Artificial bilayer lipid membranes are nonconducting either in the presence or in the absence of SNAREs (**e**, **f**). Current verses time traces of bilayer membranes containing proteins involved in docking and fusion of synaptic vesicles while the membranes are held at −60 mV (current/reference voltage). (**e**) When t-SNAREs are added to the planar lipid bilayer containing the synaptic vesicle protein, VAMP-2, no occurrence of current spike for fusion event at the bilayer membrane is observed ($n = 7$). (**f**) Similarly, no current spike is observed when t-SNAREs (syntaxin 1A-1 and SNAP25) are added to the cis side of a bilayer chamber following with VAMP-2. Increasing the concentration of t-SNAREs and VAMP-2 protein (Figure from our earlier publication: *Biophys J* 2002, 83:2522–2527). ©Bhanu Jena

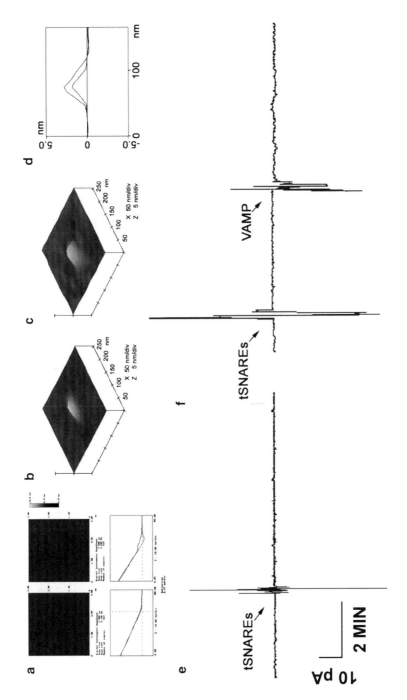

Fig. 11 AFM micrographs and force plots of mica and lipid surface and of SNAREs on lipid membrane. AFM performed on freshly cleaved mica (**a**, *left*), and on lipid membrane formed on the same mica surface (**a**, *right*), demonstrating differences in the force versus distant curves. Note the curvilinear shape exhibited in the force versus distant curves of the lipid surface in contrast to mica.

t-SNARE complexes (Fig. 11d) in lipid membrane, prior to (Fig. 11b), and following addition of v-SNARE (Fig. 11c), demonstrates changes only in the size of the complex. A 5% increase in diameter and 40% increase in height were observed following addition of v-SNARE in suspension to t-SNARE-reconstituted lipid membrane. Concomitant electrophysiological studies using the bilayer EPC9 setup demonstrated no chance in membrane conductance, supporting the AFM observations. Addition of t-SNAREs to v-SNARE-reconstituted lipid membranes did not alter membrane current (Fig. 11e). Similarly, when t-SNAREs were added to the lipid membrane prior to addition of v-SNARE, no change in the baseline current of the bilayer membrane was demonstrated (Fig. 11f). By contrast, when t-SNAREs and v-SNARE in opposing bilayers were exposed to each other, they interact and arrange in circular pattern, forming channel-like structures (Fig. 12a–d). These channels are conducting, since some vesicles are seen to have discharged their contents and are therefore flattened (Fig. 12b), measuring only 10–15 nm in height as compared with the 40–60-nm size of filled vesicles (Fig. 12a). Since the t-/v-SNARE complex lies between the opposing bilayers, the discharged vesicles clearly reveal the t-/v-SNARE ring complexes with a channel at the center (Fig. 12b–d). On the contrary, unfused v-SNARE vesicles associated with the t-SNARE-reconstituted lipid membrane exhibit only the vesicle profile (Fig. 12a). These high-resolution morphological studies using AFM demonstrate that the t-/v-SNARE arrangement is in a circular array, having a channel-like opening at the center of the complex.

To further determine if the t-/v-SNARE channels were capable of establishing continuity between the opposing bilayers, changes in current across the bilayer were examined using the EPC9 electrophysiological setup. T-SNARE vesicles containing the antifungal agent nystatin and the cholesterol homologue ergosterol were added to the *cis* compartment of the bilayer chamber containing v-SNARE in the bilayer membrane. Nystatin in the presence of ergosterol forms a cation-conducting channel in lipid membranes [84, 88, 90, 91]. When vesicles containing nystatin and ergosterol incorporate into an ergosterol-free membrane, a current spike can be observed since the nystatin channel collapses as ergosterol diffuses into the lipid membrane [88, 91, 92]. As a positive control, a KCl gradient was established to test the ability of vesicles to establish continuity at the lipid membrane (410 mM *cis*:150 mM *trans*). The KCl gradient provided a driving force for vesicle fusion that is independent of the presence of SNARE proteins [88]. When t-SNARE vesicles were exposed to v-SNARE-reconstituted bilayers, vesicles established continuity with the membrane (Fig. 12e). Fusions of t-SNARE containing vesicles with the membrane were observed as current spikes. To verify if the channel-like structures were continuous across the membrane, capacitance and conductance measurements of the membrane were carried out (Fig. 13a). Phospholipid vesicles that come in contact with the bilayer membrane do not readily fuse with the membrane. When v-SNARE-reconstituted phospholipid vesicles were added to the *cis* compartment of the bilayer chamber, a small increase in capacitance and a simultaneous increase in conductance was observed with little or no further increase over a 5 min period. This initial increase, with no further change in conductance or capacitance, is consistent with vesicles making contact with the membrane but not fusing (Fig. 12b).

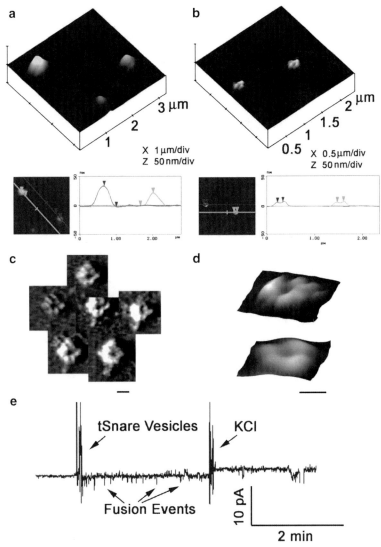

Fig. 12 Pore-like structures are formed when t-SNAREs and v-SNARE in opposing bilayers interact. (**a**) Unfused v-SNARE vesicles on t-SNARE-reconstituted lipid membrane. (**b**) Dislodgement and/ or fusion of v-SNARE-reconstituted vesicles with a t-SNARE-reconstituted lipid membrane exhibit formation of channel-like structures due to the interaction of v- and t-SNAREs in a circular array. The size of these channels range between 50 and 150 nm (**b–d**). Several 3D AFM amplitude images of SNAREs arranged in a circular array (**c**) and some at higher resolution (**d**), illustrating a channel-like structure at the center is depicted. Scale bar is 100 nm. Recombinant t-SNAREs and v-SNARE in opposing bilayers drive membrane fusion. (e) When t-SNARE vesicles were exposed to v-SNARE-reconstituted bilayers, vesicles fused. Vesicles containing nystatin/ergosterol and t-SNAREs were added to the *cis* side of the bilayer chamber. Fusion of t-SNARE containing vesicles with the membrane observed as current spikes that collapse as the nystatin spreads into the bilayer membrane. To determine membrane stability, the transmembrane gradient of KCl was increased, allowing gradient-driven fusion of nystatin-associated vesicles (Figure from our earlier publication: *Biophys J* 2002, 83:2522–2527). ©Bhanu Jena

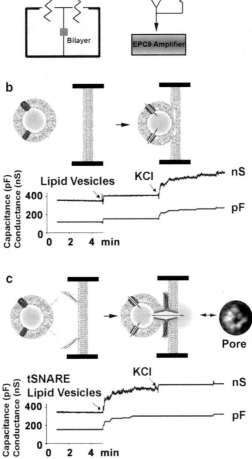

Fig. 13 Opposing bilayers containing t- and v-SNAREs interact in a circular array to form conducting pores. (**a**) Schematic diagram of the bilayer-electrophysiology setup. (**b**) Lipid vesicle containing nystatin channels (in *red*) and both vesicles and membrane bilayer without SNAREs demonstrate no significant changes in capacitance and conductance. Initial increase in conductance and capacitance are due to vesicle–membrane attachment. To demonstrate membrane stability (both bilayer membrane and vesicles), the transmembrane gradient of KCl was increased to allow gradient-driven fusion and a concomitance increase in conductance and capacitance. (**c**) When t-SNARE vesicles were added to a v-SNARE membrane support, the SNAREs in opposing bilayers arranged in a ring pattern, forming pores (as seen in the AFM micrograph on the *extreme right*) and there were seen stepwise increases in capacitance and conductance (−60 mV holding potential). Docking and fusion of the vesicle at the bilayer membrane open vesicle-associated nystatin channels and SNARE-induced pore formation, allowing conductance of ions from the *cis* to the *trans* side of the bilayer membrane. Then further addition of KCl to induce gradient-driven fusion resulted in little or no further increase in conductance and capacitance, demonstrating docked vesicles have already fused (Figure from our earlier publication: *Biophys J* 2002, 83:2522–2527). ©Bhanu Jena

These vesicles were fusogenic, since a salt (KCl) gradient across the bilayer membrane, inducing fusion of vesicles with the lipid membrane. When t-SNARE vesicles containing nystatin and ergosterol were added to the *cis* compartment of the bilayer chamber, an initial increase in capacitance and conductance occurred followed by a stepwise increase in both membrane capacitance and conductance (Fig. 13c), along with several fusion events observed as current spikes in separate recordings (Fig. 12e). The stepwise increase in capacitance demonstrates that the docked t-SNARE vesicles are continuous with the bilayer membrane. The simultaneous increase in membrane conductance is a reflection of the vesicle-associated nystatin channels that are conducting through SNARE-induced channels formed, allowing conductance of ions from the *cis* to the *trans* compartment of the bilayer chamber. SNARE-induced fusion occurs at an average rate of four t-SNARE vesicle incorporations every 5 min into the v-SNARE-reconstituted bilayer without osmotic pressure, compared with six vesicles using a KCl gradient ($n=7$). These studies demonstrate that when opposing bilayers meet, t-SNAREs in one membrane interact with v-SNAREs in the opposing membrane to form a conducting t-/v-SNARE ring complex in the presence of calcium [63].

Membrane Curvature Dictates the Size of the SNARE Ring Complex

SNARE ring complexes ranging in size from approximately 15 to 300 nm in diameter are formed when t-SNARE-reconstituted and v-SNARE-reconstituted artificial lipid vesicles meet. Since vesicle curvature would dictate the contact area between opposing vesicles, this broad spectrum of SNARE complexes observed may be due to the interaction between SNARE-reconstituted vesicles of different sizes. To test this hypothesis, t-SNARE- and v-SNARE-reconstituted proteoliposomes of distinct diameters were used [64]. Lipid vesicles of different sizes used in the study were isolated using published extrusion method [58]. The size of each vesicle population was further assessed using the AFM (Fig. 14). AFM section analysis demonstrates the presence of small 40–50 nm in diameter vesicles isolated using a 50-nm extruder filter (Fig. 14a, b). Similarly, representative samples of large vesicles measuring 150–200 and 800–1,000 nm were obtained using different size filters in the extruder. Such large vesicles are shown in the AFM micrograph (Fig. 14c, d). Analysis of vesicle size using photon correlation spectroscopy further confirmed the uniformity in the size of vesicles within each vesicle population. The morphology and size of the SNARE complex formed by the interaction of t-SNARE- and v-SNARE-reconstituted vesicles of different diameters were examined using the AFM (Fig. 15). In each case, the t-SNARE and v-SNARE proteins in opposing proteoliposomes interact and self-assemble in a circular pattern, forming channel-like structures. The interaction and arrangement of SNAREs in a characteristic ring pattern were observed for all populations of proteoliposomes examined (Fig. 15a–d). However, the size of the SNARE complex was demonstrated to be dictated by the diameter of the proteoliposomes

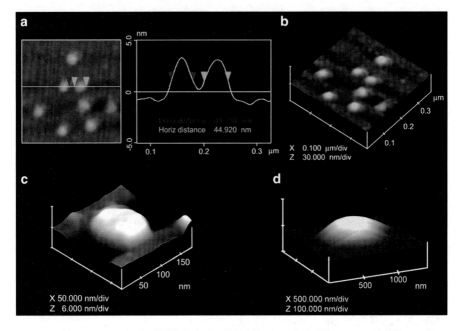

Fig. 14 AFM micrograph of t-SNARE- and v-SNARE-reconstituted liposomes of different sizes. Note the ~40–50-nm vesicles (**a**, **b**), the ~150-nm (**c**), and ~800-nm vesicle (**d**) (Figure from our earlier publication: *J Am Chem Soc* 2005, 127:10156–10157). ©Bhanu Jena

used (Fig. 15, [64]). When small (~50 nm) t-SNARE- or v-SNARE-reconstituted vesicles were allowed to interact with t-SNARE- or v-SNARE-reconstituted membrane, small SNARE-ring complexes were generated (Fig. 15a, b, [64]). With increase in the diameter of proteoliposomes, larger t-/v-SNARE complexes were formed (Fig. 15c, d). A strong linear relationship between size of the SNARE complex and vesicle diameter is further demonstrated from these studies (Fig. 16, [64]). The experimental data fit well with the high correlation coefficient, $R^2 = 0.9725$ between vesicle diameter and SNARE complex size (Fig. 16).

Disassembly of the t-/v-SNARE Complex

Studies demonstrate that the soluble N-ethylmaleimide-sensitive factor (NSF), an ATPase, disassembles the t-/v-SNARE complex in the presence of ATP [83]. This study was also the first conformation by direct physical observation that NSF–ATP is sufficient for SNARE complex disassembly. In this study, using purified recombinant NSF, and t- and v-SNARE-reconstituted liposomes, the disassembly of the t-/v-SNARE complex was examined. Lipid vesicles ranging in size from 0.2 to 2 µm were reconstituted with either t-SNAREs or v-SNARE. Kinetics of association and dissociation of t-SNARE- and v-SNARE-reconstituted liposomes in solution, in the

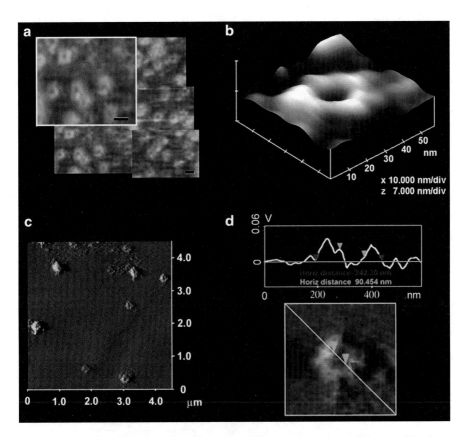

Fig. 15 Representative AFM micrograph of t-/v-SNARE complexes formed when small (**a, b**) or large (**c, d**) t-SNARE- and v-SNARE-reconstituted vesicles interact with each other. Note the formation of different size SNARE complexes, which are arranged in a ring pattern. Bar = 20 nm. AFM section analysis (**d**) shows the size of a large SNARE complex (Figure from our earlier publication: *J Am Chem Soc* 2005, 127:10156–10157). ©Bhanu Jena

presence or absence of NSF, ATP, and AMP–PNP (the nonhydrolyzable ATP analog), were monitored by right angle light scattering (Fig. 17a, b). Addition of NSF and ATP to the t/v-SNARE-vesicle mixture led to a rapid and significant increase in intensity of light scattering (Fig. 17a, b), suggesting rapid disassembly of the SNARE complex and dissociation of vesicles. Dissociation of t-/v-SNARE vesicles occurs on a logarithmic scale that can be expressed by first-order equation, with rate constant $k = 1.1$ s^{-1} (Fig. 17b). To determine whether NSF-induced dissociation of t- and v-SNARE vesicles is energy driven, experiments were performed in the presence and absence of ATP and AMP–PNP. No significant change with NSF alone, or in the presence of NSF–AMP–PNP, was observed (Fig. 17c). These results demonstrate that t-/v-SNARE disassembly is an enzymatic and energy-driven process.

To further confirm the ability of NSF–ATP in the disassembly of the t-/v-SNARE complex, immunochemical studies were performed. It has been demonstrated that

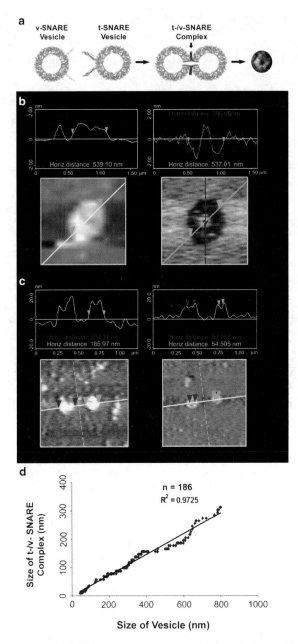

Fig. 16 SNARE complex is directly proportional to vesicle diameter. Schematic diagram depicting the interaction of t-SNARE- and v-SNARE-reconstituted vesicles. At the *extreme right*, is a single t-/v-SNARE complex imaged by AFM (**a**). AFM images of vesicles before and after their removal by the AFM cantilever tip, exposing the t-/v-SNARE complex (**b**). Interacting t-SNARE- and v-SNARE vesicles imaged by AFM at low (<200 pN) and high forces (300–500 pN). Note, at low imaging forces, only the vesicle profile is imaged (*left* **c**). However at higher forces, the soft vesicle is flattened, allowing the SNARE complex to be imaged (*right* **c**). Plot of vesicle diameter vs. size of the SNARE complex. Note the high correlation coefficient ($R^2 = 0.9725$) between vesicle diameter and the size of the SNARE complex (**d**) (Figure from our earlier publication: *J Am Chem Soc* 2005, 127:10156–10157). ©Bhanu Jena

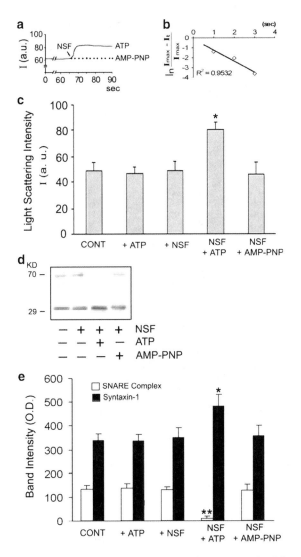

Fig. 17 NSF–ATP-induced dissociation of t-SNARE- and v-SNARE-associated liposomes. (**a**) Real-time light scattering profiles of interacting t-SNARE and v-SNARE vesicles in solution in the presence and absence of NSF (depicted by *arrow*). In the presence of ATP, NSF rapidly disassembles the SNARE complex and dissociates SNARE vesicles represented as a rapid increase in light scattering. No change in light scattering is observed when ATP is replaced with the nonhydrolyzable analog AMP–PNP. (**b**) Kinetics of NSF-induced dissociation. The graph depicts first-order kinetics of vesicles dissociation elicited by NSF–ATP. (**c**) NSF requires ATP to dissociate vesicles. NSF in the presence of ATP dissociates vesicles ($p < 0.05$, $n = 4$, Student's t test). However, NSF alone or NSF in the presence of AMP–PNP had no effect on the light scattering properties of SNARE-associated vesicle ($p > 0.05$, $n = 4$, Student's t test). (**d**) When t- and v-SNARE vesicles are mixed in the presence or absence of ATP, NSF, NSF + ATP, or NSF + AMP-PNP, and resolved by SDS–PAGE followed by immunoblots using syntaxin-1-specific antibody, t-/v-SNARE disassembly was found to be complete only in the presence of NSF–ATP (**e**). Densitometric scan of the bands reveals significant changes in SNARE complex and syntaxin-1 reactivity only when NSF and ATP were included in reaction mixture ($p < 0.05$, $n = 3$; and $p < 0.01$, $n = 3$, Student's t test) (Figure from our earlier publication: *J Am Chem Soc* 2006, 128:26–27). ©Bhanu Jena

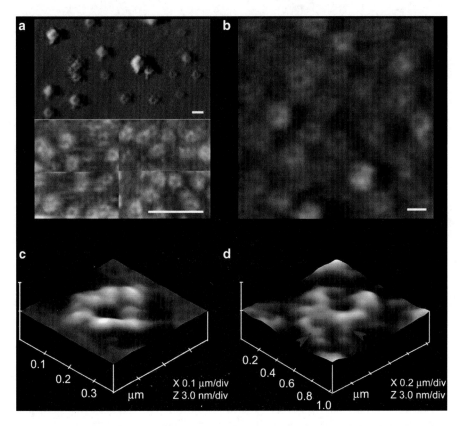

Fig. 18 AFM micrographs of NSF–ATP-induced disassembly of the t-/v-SNARE ring complex. Representative AFM micrograph of t-/v-SNARE complexes formed when large (*top panel* **a**) or small (*bottom panel* **a**) t-/v-SNARE ring complexes are formed due to the interaction of large and small v-SNARE-reconstituted vesicles interact with a t-SNARE-reconstituted lipid membrane. Bar = 250 nm. (**b**) Disassembly of large t-/v-SNARE complex. Bar = 250 nm. (**c**) High resolution of a t-/v-SNARE ring complex, and a disassembled one (**d**) (Figure from our earlier publication: *J Am Chem Soc* 2006, 128:26–27). ©Bhanu Jena

v-SNARE and t-SNAREs form an SDS-resistant complex [10]. NSF binds to SNAREs and forms a stable complex when locked in the ATP-bound state (ATP–NSF). Thus, in the presence of ATP + EDTA, VAMP antibody has been demonstrated to be able to co-immunoprecipitate this stable NSF–SNARE complex [93]. Therefore, when t- and v-SNARE vesicles were mixed in the presence or absence of ATP, NSF, NSF + ATP, or NSF + AMP–PNP, and resolved using SDS–PAGE followed by immunoblot analysis using syntaxin-1-specific antibody, t-/v-SNARE disassembly was found to be complete only in the presence of NSF–ATP (Fig. 17d, e). To further confirm these findings (Fig. 17), direct observation of the t-/v-SNARE complex disassembly was assessed using AFM. When purified recombinant t-SNAREs and v-SNARE in opposing bilayers interact and self-assemble to form supramolecular ring complexes, they disassembled when exposed to recombinant

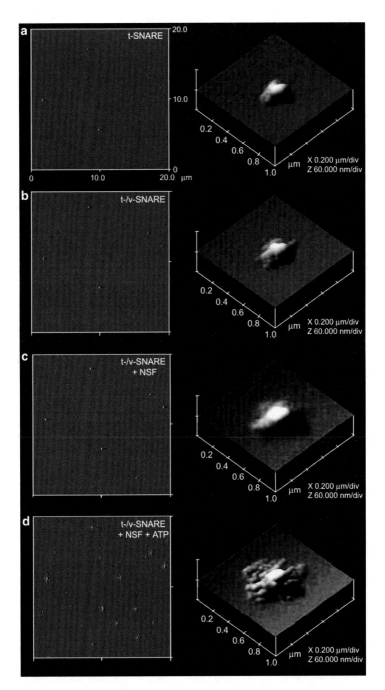

Fig. 19 AFM micrographs of NSF–ATP-induced disassembly of the t-/v-SNARE complex formed when v-SNARE is added to a t-SNARE-reconstituted lipid membrane. The *left panel* (**a–d**) shows at low resolution, the sequential AFM micrographs of one of the ten representative experiments, where v-SNARE is added to a t-SNARE-reconstituted lipid membrane, followed by NSF and then ATP. Note the dramatic disassembly of the SNARE complexes in (**d**). The *right panel* shows at higher resolution, the disassembly of one of such SNARE complexes (Figure from our earlier publication: *J Am Chem Soc* 2006, 128:26–27). ©Bhanu Jena

NSF and ATP, as observed at nanometer resolution using AFM (Fig. 18). Since SNARE ring complex requires v-SNARE and t-SNAREs to be membrane associated, suggested that NSF may require the t-/v-SNARE complex to be arranged in a specific configuration or pattern, for it to bind and disassemble the complex in the presence of ATP. To test this hypothesis, t-SNAREs followed by v-SNARE, NSF, and ATP were added to a lipid membrane and continuously imaged in buffer by AFM (Fig. 19). Results from this study demonstrate that both SNARE complexes either in the presence or in the absence of membrane disassemble [83]. Furthermore, close examination using AFM, the NSF–ATP-induced disassembly of SNARE complex, demonstrates NSF to function as a right-handed molecular motor [82].

CD Spectroscopy Confirms Membrane Requirement for Appropriate t-/v-SNARE Assembly and Sufficiency of NSF–ATP for Its Disassembly

The overall secondary structural content of full-length neuronal v-SNARE and t-SNAREs, and the t-/v-SNARE complex, both in suspension and in association with membrane, has been determined by CD spectroscopy using an Olis DSM 17 spectrometer [89]. Circular dichroism spectroscopy reveals that v-SNARE in buffered suspension (Fig. 20ai), when incorporated into liposomes (Fig. 20bi), exhibits reduced folding (Table 1). This loss of secondary structure following incorporation of full-length v-SNARE in membrane may be a result of self-association of the hydrophobic regions of the protein in the absence of membrane. When incorporated into liposomes, v-SNARE may freely unfold without the artifactual induction of secondary structure, as reflective of the lack in CD signals at 208 and 222 nm, distinct for α-helical content. The t-SNAREs (Fig. 20aii, bii) show clearly defined peaks at both these wavelengths, consistent with a higher degree of helical secondary structures formed both in buffered suspension and in membrane, at ca. 66% and 20%, respectively (Table 1). Again, the membrane-associated SNAREs exhibit less helical content than when in suspension. Similarly, there appears to be a dramatic difference in the CD signal observed in t-/v-SNARE complexes in suspension, and those complexes that are formed when membrane-associated SNAREs interact (Fig. 20aiii, biii). Interestingly, there is no increase in secondary structure upon complex formation. Rather, the CD spectra of the complexes are identical to a combination of individual spectra. Moreover, membrane-associated t-/v-SNAREs are less folded than the purified SNARE complex. This data supports previous AFM results that lipid is required for proper arrangement of the SNARE proteins in membrane fusion. Addition of NSF to the t-/v-SNARE complex results in an increase in the unordered fraction (Fig. 20aiv, biv; Table 1), which may be attributed to an overall disordered secondary structure of the NSF, and not necessarily unfolding of the t-/v-SNARE complex. By contrast, activation of NSF by

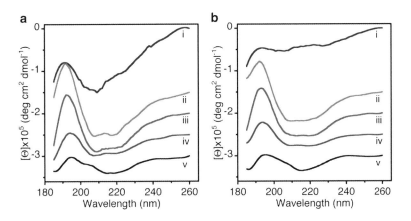

Fig. 20 Circular dichroism data reflecting structural changes to SNAREs, both in suspension and in association with membrane. Structural changes, following the assembly and disassembly of the t-/v-SNARE complex is further shown. (**a**) CD spectra of purified full-length SNARE proteins in suspension and (**b**) in association with membrane; their assembly and NSF–ATP-induced disassembly are demonstrated. (*i*) v-SNARE; (*ii*) t-SNAREs; (*iii*) t-/v-SNARE complex; (*iv*) t-/v-SNARE + NSF, and (*v*) t-/v-SNARE + NSF + 2.5 mM ATP are shown. CD spectra were recorded at 25°C in 5 mM sodium phosphate buffer (pH 7.5) at a protein concentration of 10 μM. In each experiment, 30 scans were averaged per sample for enhanced signal to noise, and data were acquired on duplicate independent samples to ensure reproducibility (Figure from our earlier publication: *Chem Phys Lett* 2008, 462:6–9). ©Bhanu Jena

the addition of ATP almost completely abolishes all α-helical content within the multiprotein complex (Fig. 20av, bv). This direct observation of the helical unfolding of the SNARE complex using CD spectroscopy under physiologically relevant conditions (i.e., in membrane-associated SNAREs) confirms earlier AFM reports on NSF–ATP-induced t-/v-SNARE complex disassembly [83]. In further agreement with previously reported studies using the AFM, the consequence of ATP addition to the t-/v-SNARE–NSF complex is disassembly, regardless of whether the t-/v-SNARE + NSF complex is membrane associated or in buffered suspension. These studies further demonstrate that higher SNARE protein concentrations are without influence on the membrane-directed self-assembly of the SNARE complex [89]. In summary, the CD results demonstrate that v-SNARE in suspension, when incorporated into liposomes, exhibits reduced folding. Similarly, t-SNAREs which exhibit clearly defined peaks at CD signals of 208 and 222 nm wavelengths, consistent with a higher degree of helical secondary structure in both the soluble and liposome-associated forms, exhibit reduced folding when membrane associated. ATP-induced activation of NSF bound to the t-/v-SNARE complex results in disassembly of the SNARE complex, eliminating all α-helices within the structure. In addition, these studies are a further confirmation of earlier reports [83] that NSF–ATP is sufficient for the disassembly of the t-/v-SNARE complex.

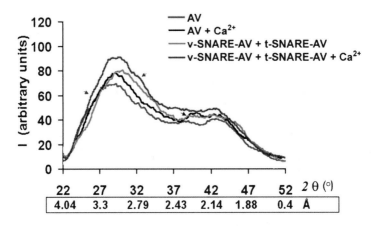

Fig. 21 Wide-angle X-ray diffraction patterns on interacting lipid vesicles. Representative diffraction profiles from one of the four separate experiments using plain and t- and v-SNARE-reconstituted lipid vesicles, both in the presence or in the absence of 5 mM Ca²⁺, are shown. Note the shift in the major peak to the right, when t-SNARE- and v-SNARE-reconstituted vesicles interact (Figure from our earlier publication: *Cell Biol Int* 2004, 28:19–31). ©Bhanu Jena

SNAREs Bring Opposing Bilayers into Close Approximation, Enabling Calcium Bridging and Membrane Fusion

Diffraction patterns of nonreconstituted vesicles and t- and v-SNARE-reconstituted vesicles in the absence and presence of 5 mM Ca²⁺ are shown for comparison in Fig. 21. To our knowledge, these are the first recorded wide-angle diffractograms of unilamellar vesicles, in the 2–4 Å diffraction range [58]. They have broad pattern spanning 2θ which ranges approximately 23–48° or d values of 3.9–1.9 Å with sharp drop off intensity on either side of the range. Relatively, broad feature of diffractogram indicate multitude of contacts between atoms of one vesicle as well between different vesicles during collision. However, two broad peaks are visible on the diffractogram, the stronger one at 3.1 Å and a weaker one at 1.9 Å. They indicate that the greatest number of contacts between them have these two distances. Addition of Ca²⁺ or incorporation of SNAREs at the vesicles membrane or both influence both peaks within the 2.1–3.3 Å intensity range (Fig. 21). However, the influence of Ca²⁺, SNAREs, or both is more visible on peak positioned at 3.1 Å in form of an increased I_{max} of arbitrary units and 2θ. This increase in I_{max} at the 3.1 Å can be explained in terms of increased vesicle pairing and/or a decrease in distance between apposed vesicles. Incorporation of t- and v-SNARE proteins at the vesicle membrane allows for tight vesicle–vesicle interaction, which is demonstrated again as an I_{max} shifts to 30.5 or 2.9 Å from 3.1 Å. Ca²⁺ and SNAREs work in manner that induces a much higher increase in peak intensity with appearance of shoulders on both sides of the peak at 2.8 and 3.4 Å (Fig. 21). This indicates an increase in number of vesicle contact points at a constant distance between them. Vesicles containing either t- or v-SNAREs have little effect on the X-ray scattering patterns. Only as

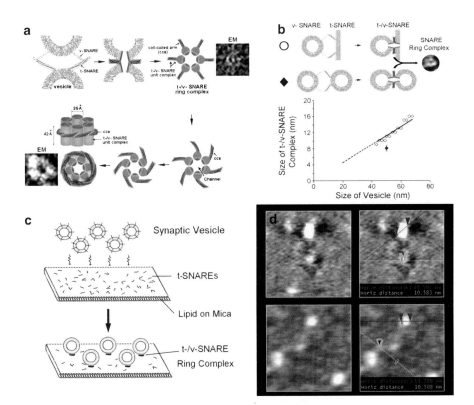

Fig. 22 The possible establishment of a leak-proof SNARE ring complex channel is demonstrated (**a**). Size of the t-/v-SNARE ring complex is directly proportional to the size of the SNARE-associated vesicle (**b**). Different sizes of v-SNARE-associated vesicles, when interact with t-SNARE-associated membrane (*white circle*), demonstrate the SNARE ring size to be directly proportional to the vesicle size. When a 50 nm in diameter v-SNARE-reconstituted vesicle interacts with a t-SNARE-reconstituted membrane, an 11 nm in diameter t-/v-SNARE ring complex is formed. Similarly, the present study demonstrates that when a 50 nm in diameter v-SNARE-reconstituted vesicle interacts with a 50 nm in diameter t-SNARE-reconstituted vesicle, a 8 nm in diameter t-/v-SNARE ring complex is established (*black diamond*). Analogous to the 11 nm in diameter t-/v-SNARE ring complexes formed when 50 nm v-SNARE vesicles meet a t-SNARE-reconstituted planer membrane (**b**), approximately 11 nm in diameter t-/v-SNARE ring complexes are formed when 50 nm in diameter synaptic vesicles meet a t-SNARE-reconstituted planer membrane (**c, d**) (Figure from our earlier publication: *J Cell Mol Med* 2011, 15:31–37). ©Bhanu Jena

discussed above, when t-SNARE- and v-SNARE-reconstituted vesicles were brought together, did we detect a change in the X-ray diffraction pattern. Since exposure of t-SNARE vesicle and v-SNARE vesicle mixture to Ca^{2+} results in maximum increase in a.u. and 2θ using X-ray diffraction, the effect of Ca^{2+} on fusion and aggregation of t-/v-SNARE vesicles were examined using light scattering, light microscopy, AFM, fluorescent dequenching, and electrical measurements of fusion [58].

In recent studies [65], using high-resolution electron microscopy, the electron density maps and 3D topography of the membrane-directed SNARE ring complex were determined at nanometer resolution (Fig. 22). Similar to the t-/v-SNARE ring complex formed when 50-nm v-SNARE liposomes meet a t-SNARE-reconstituted planer membrane, SNARE rings are also formed when 50 nm in diameter isolated synaptic vesicles meet a t-SNARE-reconstituted planer lipid membrane. Furthermore, the mathematical prediction of the SNARE ring complex size with reasonable accuracy and the possible mechanism of membrane-directed t-/v-SNARE ring complex assembly were determined from the study. Using both liposome-reconstituted recombinant t-/v-SNARE proteins and native v-SNARE present in isolated synaptic vesicle membrane, the membrane-directed molecular assembly of the neuronal SNARE complex was revealed for the first time and its size mathematically predicted [65]. These results provide a new molecular understanding of the universal machinery and mechanism of membrane fusion in cells, having fundamental implications in human health and disease.

Membrane Lipids Influence SNARE Complex and Assembly–Disassembly

Cholesterol and lysophosphatidylcholine (LPC) are known to contribute to the negative and positive curvature, respectively, of membranes [94, 95]. Studies demonstrate [96] that membrane-containing LPC generates larger SNARE ring complexes (Fig. 23), where the α-helical component of the complex is little affected by NSF–ATP [96]. By contrast, cholesterol-containing membrane produces smaller SNARE ring complexes that readily disassemble in the presence of NSF–ATP (Fig. 23, [96]). Using CD spectroscopy, SNARE ring complexes formed in the presence of either cholesterol or LPC demonstrate profound differences. As previously demonstrated [89], high α-helical content in t-SNARE and t-/v-SNARE complexes are present in both. However, in the presence of NSF and ATP, peaks at 208 and 222 nm, characteristic of α-helical secondary structure, are abolished in the cholesterol groups. CD spectrographs on membrane-associated v-SNARE display little signal, as previously demonstrated [89]; however, v-SNARE reconstituted in liposomes containing cholesterol displays CD signals for α-helical content. By contrast, the LPC groups exhibit no signal for α-helical content. This was the first report that membrane curvature-influencing lipids profoundly influence SNARE complex size and its disassembly [96]. As previously reported [89] in membrane containing no cholesterol or LPC, NSF–ATP induces disassembly of the α-helical contents, not the β-sheet structures in the t-/v-SNARE complex. By contrast, in the presence of LPC, NSF–ATP induces disassembly of the β-sheet structures, and not the α-helical structures within the SNARE complex. Studies implicate cholesterol's role in membrane fusion to be indirect, centered on SNARE formation through cholesterol binding to synaptophysin, a calcium- and cholesterol-dependent vesicle-associated protein which forms a complex with synaptobrevin (VAMP), subsequently facilitating v-SNARE interaction with t-SNAREs [97]. In the in vitro cholesterol/LPC study [96] however, no synaptophysin is present to influence such interactions of cholesterol with SNAREs, and therefore little or no

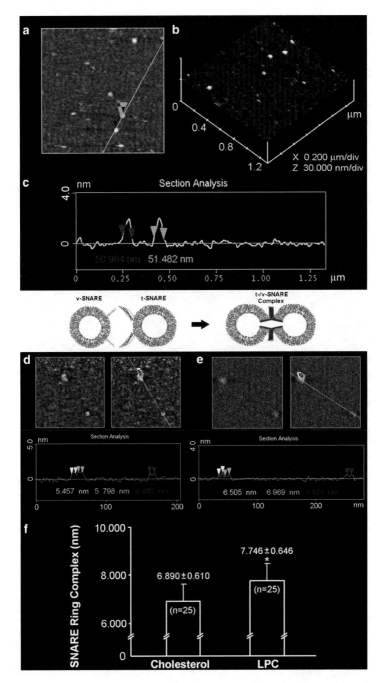

Fig. 23 Representative AFM micrographs of approximately 50 nm in diameter liposomes and the t-/v-SNARE ring complexes formed when such cholesterol or LPC containing t-SNARE and v-SNARE proteoliposomes meet [96]. Note the 50–53 nm in diameter cholesterol-containing liposomes (**a–c**). Similar size LPC-containing vesicles were prepared and observed using the AFM (data not shown). Note the 6.89 ± 0.61 nm t-/v-SNARE ring complexes formed when approximately 50 nm in diameter t-SNARE–cholesterol–liposomes interact with 50 nm v-SNARE–cholesterol vesicles (**d**, **f**). Similarly, 7.746 ± 0.646 nm t-/v-SNARE ring complexes are formed when cholesterol is replaced by LPC (**e**, **f**).(*$p < 0.001$). (Figure from our earlier publication: *J Am Chem Soc* 2010, 132:5596–5597). ©Bhanu Jena

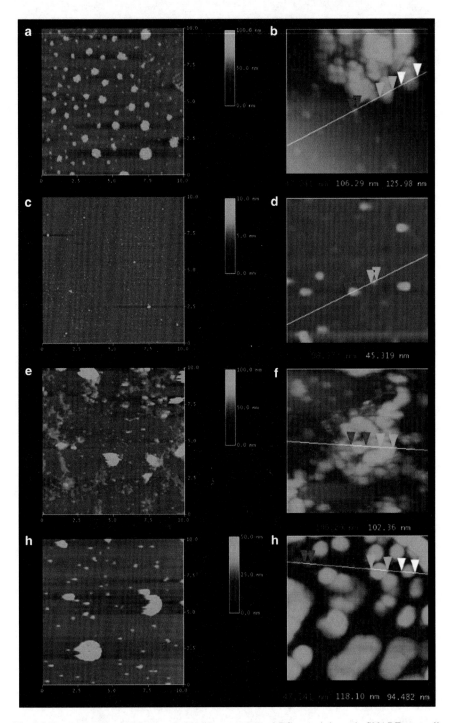

Fig. 24 Representative AFM micrographs demonstrating LPC containing t-/v-SNARE proteoliposome complexes fail to dissociate in the presence of NSF–ATP. Exposure of cholesterol-associated t-SNARE and v-SNARE liposome mixtures (**a, b**, low and high magnification) to NSF–ATP results in liposome dissociation as demonstrated in (**c**) at low magnification and (**d**) at higher magnification. By contrast, LPC-associated t-/v-SNARE liposomes (**e, f**) remain clustered (**g, h**) following exposure to NSF–ATP (Figure from our earlier publication: *J Cell Mol Med* 2011, doi:10.1111/j.1582-4934.2011.01433.x). ©Bhanu Jena

effect of cholesterol is demonstrated on the α-helical and β-sheet content of membrane-associated SNAREs and the SNARE complex. The influence of LPC on SNARE assembly–disassembly has been further tested using AFM and X-ray diffraction studies (Fig. 24). Two sets of 50 nm in diameter liposomes, one set containing cholesterol and the other LPC, were reconstituted with either t-SNAREs or v-SNARE for use. Exposure of cholesterol-associated t-SNARE and v-SNARE liposome mixtures resulted in the formation of vesicle clusters due to the interaction of t-SNARE in one vesicle interacting with v-SNARE in the opposing vesicle, as observed using AFM (Fig. 24a, b). Exposure of the vesicle clusters to NSF–ATP resulted in dissociation of the clusters due to NSF–ATP-induced t-/v-SNARE complex disassembly (Fig. 24c, d). The presence of vesicles as monomers or dimers is observed following t-/v-SNARE disassembly (Fig. 24d). By contrast, exposure of LPC-associated t-SNARE and v-SNARE liposome mixtures to NSF–ATP resulted in little or no NSF–ATP-induced disassembly, and consequently the accumulation of vesicles in clusters (Fig. 24e–h).

To further determine the influence of LPC and cholesterol on the interaction between t-SNARE and v-SNARE vesicles, X-ray diffraction studies have been carried out. Recordings made of vesicles in solution in the 1.54–5.9 Å diffraction range exhibit the characteristic broad diffraction pattern, spanning 2θ ranges 26.67–42.45° or d values of 2.1–3.3 Å. The diffractogram trace exhibits a pattern typical of short-range ordering in a liquid system, indicating a multitude of contacts between interacting vesicles, the majority being in the 3 Å region. In agreement with AFM studies, X-ray studies demonstrate larger clusters and consequently much less diffraction by the LPC vesicles compared with cholesterol. The distance however between vesicles is determined to be closer in the cholesterol population (3.05 Å) than in the LPC group (3.33 Å).

These findings support the existence of direct lipid–protein interactions to differentially modulate SNARE function within various cellular compartments. Modulating the concentration and distribution of such nonbilayer lipids at various membranes could regulate the efficacy and potency of membrane fusion and membrane-directed SNARE complex assembly–disassembly in cells. Cells with higher membrane cholesterol levels, would promote membrane fusion, while cells with increased membrane LPC content would facilitate secretory event longevity by inhibiting SNARE complex disassembly. This hypothesis has recently been tested in live cells ([98], Fig. 25).

To determine the role of cholesterol and LPC on t-/v-SNARE complex disassembly in live cells, studies have been carried out in brain and pancreatic tissue preparations (Fig. 25). Isolated rat brain slices exposed to cholesterol, LPC, or vehicle (control) were used in the study. The brain tissue was stimulated using 30 mM KCl, followed by immunoblot analysis. Since v-SNARE and t-SNAREs form an SDS-resistant complex, and NSF binds to the complex in the presence of ATP (NSF–ATP) [93], both control and experimental brains were solubilized in the presence of ATP–EDTA, and a SNARE antibody used to immunoisolate the stable NSF–t-/v-SNARE complex [93]. Results from the study demonstrate that in stimulated brain slices, inhibition of t-/v-SNARE complex disassembly in the presence of LPC is observed, further confirming AFM, X-ray, and DLS experiments (Fig. 25d).

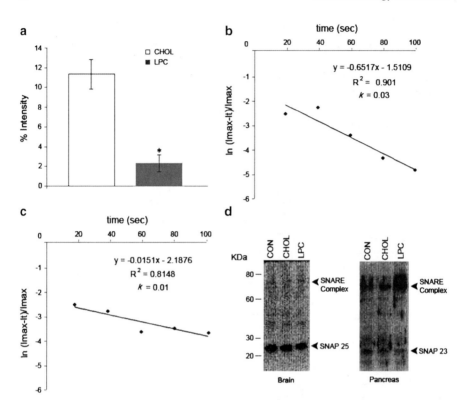

Fig. 25 SNARE complexes in the presence of LPC fail to disassemble. (**a**) Real-time dynamic light scattering (DLS) profiles on cholesterol-associated (CHOL) and LPC-associated t-/v-SNARE liposomes in the presence of NSF–ATP. There is no appreciable dissociation of the LPC vesicles in contrast to a rapid ATP-dependent dissociation of CHOL vesicles ($p < 0.001$). (**b**) Note the dissociation of cholesterol-associated t-/v-SNARE vesicles occurs with rate constant $k = 0.03$ s^{-1}, and (**c**) a relatively slow dissociation in LPC vesicles ($k = 0.01$ s^{-1}). (**d**) Following KCl stimulation, isolated brain slices preincubated in CHOL, LPC, or vehicle (CON) are solubilized in buffer containing ATP–EDTA, and 10 µg of protein resolved by SDS–PAGE, followed by immunoblot analysis using SNAP-25-specific antibody; negligible disassembly of the t-/v-SNARE complex is demonstrated in brain tissue preincubated in LPC, as opposed to the control (CON) or CHOL. Similarly exocrine pancreas preincubated in LPC demonstrates reduced disassembly of the t-/v-SNARE complex following stimulation of secretion using 1 µM carbamylcholine (Figure from our earlier publication: *J Cell Mol Med* 2011, doi:10.1111/j.1582-4934.2011.01433.x). ©Bhanu Jena

These results suggest that in the presence of LPC, once v-SNARE-associated secretory vesicles interact with t-SNARE membrane to form the t-/v-SNARE ring complex and establish continuity, the complex fails to disassemble, resulting in an inhibition of subsequent rounds of vesicle docking and fusion. To further test this hypothesis in live cells, the experiment was repeated using exocrine pancreas. Following incubation in LPC, cholesterol, or vehicle, when stimulated exocrine pancreas were examined by immunoblot analysis using SNAP-23-specific antibody (Fig. 25d), results confirmed the inhibitory effect of LPC on SNARE complex disassembly also in live pancreatic acinar cells. These results are in agreement with earlier findings supporting LPC to be a membrane fusion inhibitor [99].

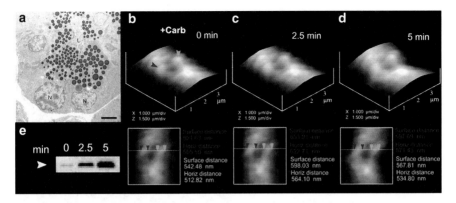

Fig. 26 The swelling dynamics of ZGs in live pancreatic acinar cells. (**a**) Electron micrograph of pancreatic acinar cells showing the basolaterally located nucleus (N) and the apically located ZGs. The apical end of the cell faces the acinar lumen (L). Bar = 2.5 μm. (**b–d**) The apical ends of live pancreatic acinar cells were imaged by AFM, showing ZGs (*red* and *green arrowheads*) lying just below the apical plasma membrane. Exposure of the cell to a secretory stimulus using 1 μM carbamylcholine resulted in ZG swelling within 2.5 min, followed by a decrease in ZG size after 5 min. The decrease in size of ZGs after 5 min is due to the release of secretory products such as α-amylase, as demonstrated by the immunoblot assay (**e**) (Figure from our earlier publication: *Cell Biol Int* 2004, 28:709–716). ©Bhanu Jena

Secretion Involves Vesicle Swelling and Content Expulsion

Secretory Vesicle Swelling Is Required for the Expulsion of Intravesicular Contents During Cell Secretion

In the past 15 years, the dynamics of intracellular membrane-bound secretory vesicles ranging in size from 200–1,200 nm in pancreatic acinar cells to 35–50 nm in neurons have been extensively studied, shedding light on the molecular process involved in vesicular discharge during cell secretion [68–70, 100–104]. Live pancreatic acinar cells in near physiological buffer, when imaged using the AFM, reveal at nanometer resolution the size of ZG lying immediately below the surface of the apical plasma membrane [68]. Within 2.5 min of exposure to a secretory stimulus, a majority of ZGs within cells swell, followed by a decrease in ZG size, and a concomitant release of secretory products [68]. These studies directly demonstrated intracellular swelling of secretory vesicles following stimulation of cell secretion in live cells, and vesicle deflation following partial discharge of vesicular contents. Furthermore, a direct estimation of vesicle size dynamics at nanometer resolution under various experimental conditions has enabled the determination of the molecular mechanism of secretory vesicle swelling [69, 70, 100–104]. AFM and PCS have been major players in these studies.

When live pancreatic acinar cells (Fig. 26) in near physiological buffer are imaged using the AFM at high force (200–300 pN), ZGs lying immediately beneath the apical plasma membrane of the cell (Fig. 26b) are observed. Within 2.5 min of exposure to a physiological secretory stimulus (1 μM carbamylcholine), a majority of ZGs within

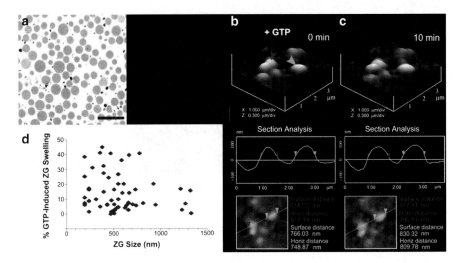

Fig. 27 Swelling of isolated ZGs. (**a**) Electron micrograph of isolated ZGs demonstrating a homogeneous preparation. Bar = 2.5 μm. (**b, c**) Isolated ZGs, on exposure to 20 μM GTP, swell rapidly. Note the enlargement of ZGs as determined by AFM section analysis of two vesicles (*red* and *green arrowheads*). (**d**) Percent ZG volume increase in response to 20 μM GTP. Note how different ZGs respond to the GTP-induced swelling differently (Figure from our earlier publication: *Cell Biol Int* 2004, 28:709–716). ©Bhanu Jena

cells swell (Fig. 26c), followed by a decrease in ZG size (Fig. 26d) by which time most of the release of secretory products from within ZGs have occurred (Fig. 26e). These studies reveal in live cells intracellular swelling of secretory vesicles following stimulation of secretion and their deflation following partial discharge of vesicular contents. Measurement of intracellular ZG size further demonstrates that different vesicles swell differently following a secretory stimulus. For example in Fig. 26b, c, the ZG marked by the red arrowhead demonstrated a 23–25% increase in diameter, in contrast to the ZG labeled green, with only 10–11% increase. This differential swelling among ZGs within the same cell may explain why following stimulation of cell secretion, some intracellular ZGs demonstrate the presence of less vesicular content than others and hence have discharged more of their contents [105]. To determine precisely the role of secretory vesicle swelling in vesicle–plasma membrane fusion and in the expulsion of intravesicular contents, an electrophysiological porosome-reconstituted lipid bilayer fusion assay [14, 68] has been employed. The ZGs used in the bilayer fusion assays are characterized for their purity and their ability to respond to a swelling stimulus. ZGs are isolated using a published procedure [69] and their purity assessed using electron microscopy (Fig. 27a). As previously reported [69, 100, 104], exposure of isolated ZGs (Fig. 27b) to GTP resulted in ZG swelling (Fig. 27c). Once again, similar to what is observed in live acinar cells (Fig. 26), each isolated ZG responds differently to the same swelling stimulus (Fig. 27). For example, the red arrowhead points to a ZG whose diameter increased by 29% as opposed to the green arrowhead pointing ZG that increased only by a modest 8% (Fig. 27b). The differential response of isolated ZGs to GTP has been further assessed by measuring changes in the volume of isolated ZGs of different sizes (Fig. 27d). ZGs in the exocrine pancreas range in size from 0.2 to 1.2 μm in diameter [69]. Not all ZGs are found to swell following a GTP challenge. Most ZG volume increases were between 5% and 20%. However, larger increases of up to 45% are observed in vesicles ranging in size from 250 to 750 nm in diameter (Fig. 27d). These studies demonstrate that following stimulation of secretion, ZGs within pancreatic acinar cells swell, followed by a release of intravesicular contents through porosomes [6, 7, 13, 14] at the cell plasma membrane, and a return to resting size on completion of secretion. On the contrary, isolated ZGs stay swollen following exposure to GTP/Mas, since there is no release of the intravesicular contents. In pancreatic acinar cells, little secretion is detected 2.5 min following stimulation of cell secretion, although the ZGs within them are fully swollen (Fig. 26c). However, at 5 min following stimulation, ZGs deflated and the intravesicular α-amylase is released from the acinar cell (Fig. 26e), suggesting the involvement of ZG swelling in cell secretion.

In the electrophysiological bilayer fusion assay (Fig. 28, [68]), immunoisolated porosomes from the exocrine pancreas are isolated and functionally reconstituted [14] into the lipid membrane of the bilayer apparatus where membrane conductance and capacitance are continually monitored (Fig. 28a). Reconstitution of the porosome into the lipid membrane results in a small increase in capacitance (Fig. 28b), due to increase in membrane surface area contributed by the incorporation of porosomes ranging in size from 100 to 150 nm in diameter [14]. In the presence of calcium, when isolated ZGs are added to the *cis* compartment of the bilayer chamber, they dock and fuse at the porosome-reconstituted lipid membrane (Fig. 28a), detected as a step increase in membrane capacitance (Fig. 28b). However, even after 15 min of ZG addition to the *cis* compartment of the bilayer chamber, little or no release of the intravesicular enzyme α-amylase is detected in the *trans* compartment of the chamber (Fig. 28c, d). On the contrary, exposure of ZGs to 20 μM GTP induced swelling [69, 101, 105] and results both in the potentiation of fusion and in a robust expulsion of α-amylase into the *trans* compartment of the bilayer chamber (Fig. 28c, d). These studies demonstrate that during cell secretion, secretory vesicle swelling is required for the efficient and precisely regulated expulsion of intravesicular contents. Within minutes or even seconds following a secretory stimulus, empty and partially empty secretory vesicles accumulate within cells [105–108]. There may be two possible explanations for such accumulation of partially empty vesicles. Following fusion at the porosome, secretory vesicles may either remain fused for a brief period and therefore time would be the limiting factor for partial expulsion, or inadequately swell and therefore unable to generate the required intravesicular pressure for complete vesicular discharge. Data in Fig. 28 suggest that it would be highly unlikely that generation of partially empty vesicles results from brief periods of vesicle fusion at porosomes. Following addition of ZGs to the *cis* compartment of the bilayer apparatus, membrane capacitance continues to increase; however, little or no detectable secretion occurs even after 15 min (Fig. 28), suggesting that either variable degrees of vesicle swelling or repetitive cycles of fusion and swelling of the same vesicle or both may operate during secretion. Under

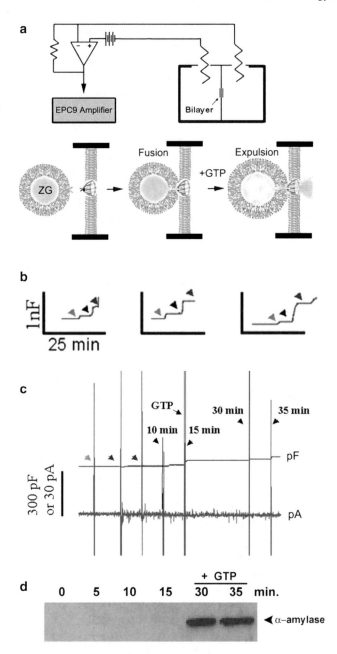

Fig. 28 Fusion of isolated ZGs at porosome-reconstituted bilayer and GTP-induced expulsion of α-amylase. (**a**) Schematic diagram of the EPC9 bilayer apparatus showing the *cis* and *trans* chambers. Isolated ZGs when added to the *cis* chamber fuse at the bilayer-reconstituted porosome. Addition of GTP to the *cis* chamber induces ZG swelling and expulsion of its contents such as α-amylase to the *trans* bilayer chamber. (**b**) Capacitance traces of the lipid bilayer from three separate experiments following reconstitution of porosomes (*green arrowhead*), addition of ZGs to the

these circumstances, empty and partially empty vesicles could be generated within cells following secretion. To test this hypothesis, two key parameters have been examined [68]. One, whether the extent of swelling is same for all ZGs exposed to a certain concentration of GTP, and two, whether ZG are capable of swelling to different degrees. And if they are capable of differential swelling, whether there is a correlation between extent of swelling and the quantity of intravesicular contents expelled. The answer to the first question is: different ZGs respond to the same stimulus differently (Fig. 26). It is demonstrated [68] that different ZGs within cells or in isolation undergo different degrees of swelling even though they are exposed to the same stimulus: carbamylcholine for live pancreatic acinar cells, and GTP for isolated ZGs (Figs. 26b–d and 27b–d). The requirement of ZG swelling for expulsion of vesicular contents has been further confirmed when a GTP dose-dependent increase in ZG swelling (Fig. 29a–c) is translated to a concomitant increase in secretion of α-amylase (Fig. 29d). Although higher GTP concentrations elicit increased ZG swelling, the extent of swelling between ZGs varies.

To determine if a similar or an alternate mechanism is responsible for the release of secretory products in a fast secretory cell, the neuron, synaptosomes, and synaptic vesicles from rat brain have been utilized in studies [68]. Since the synaptic vesicle membrane is known to possess the heterotrimeric G_o protein, we hypothesized GTP and the GTP-binding G-protein agonist mastoparan (Mas) to mediated vesicle swelling. To test this hypothesis, isolated synaptosomes (Fig. 30a) were lysed to obtain synaptic vesicles and synaptosomal membrane preparations. Isolated synaptosomal membrane when placed on mica and imaged by AFM in near physiological buffer reveals on its cytosolic compartment the presence of 35–50 nm in diameter synaptic vesicles docked to porosomes at the presynaptic membrane (Fig. 30b–d). Analogous to ZGs, exposure of synaptic vesicles (Fig. 30b) to 20 μM GTP (Fig. 30c), results in swelling and a concomitant increase in synaptic vesicle size. Exposure to Ca^{2+} results in the transient fusion of synaptic vesicles at the porosome in the presynaptic membrane, resulting in expulsion of intravesicular contents, and a consequent decrease in synaptic vesicle size (Fig. 30d, e). In Fig. 30b–d, the blue arrowhead points to a synaptic vesicle undergoing such a process. Additionally, as observed in ZGs of the exocrine pancreas, not all synaptic vesicles swell, and if they do, they exhibit differential swelling even though they have been exposed to the same stimulus. This differential response of synaptic ves-

---

Fig. 28 (continued) *cis* chamber (*blue arrowhead*), and the *red arrowhead* represents the 5-min time point after ZG addition. Note the small increase in membrane capacitance following porosome reconstitution, and a greater increase when ZGs fuse at the bilayers. (**c**) In a separate experiment, 15 min after addition of ZGs to the *cis* chamber, 20 μM GTP was introduced. Note the increase in capacitance, demonstrating potentiation of ZG fusion. Flickers in current trace represent current activity. (**d**) Immunoblot analysis of α-amylase in the *trans* chamber fluid at different times following exposure to ZGs and GTP. Note the undetectable levels of α-amylase even up to 15 min following ZG fusion at the bilayer. However, following exposure to GTP, significant amounts of α-amylase from within ZGs were expelled into the *trans* bilayer chamber (Figure from our earlier publication: *Cell Biol Int* 2004, 28:709–716). ©Bhanu Jena

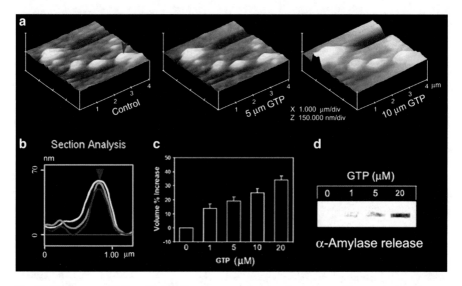

Fig. 29 The extent of ZG swelling is directly proportional to the amount of intravesicular contents released. (**a**) AFM micrographs showing the GTP dose-dependent increase in swelling of isolated ZGs. (**b**) Note the AFM section analysis of a single ZG (*red arrowhead*), showing the height and relative width at resting (control, *red outline*), following exposure to 5 μM GTP (*green outline*) and 10 μM GTP (*white outline*). (**c**) Graph demonstrating the GTP dose-dependent percent increase in ZG volume. Data are expressed as mean ± SEM. (**d**) Immunoblot analysis of α-amylase in the *trans* chamber fluid of the bilayer chamber following exposure to different doses of GTP. Note the GTP dose-dependent increase in α-amylase release from within ZGs fused at the *cis* side of the reconstituted bilayer (Figure from our earlier publication: *Cell Biol Int* 2004, 28:709–716). ©Bhanu Jena

icles within the same nerve ending may dictate and regulate the potency and efficacy of neurotransmitter release at the nerve terminal. To further confirm synaptic vesicle swelling and determine the swelling rate, dynamic light scattering and photon correlation spectroscopy experiments have been carried out. Light scattering studies demonstrate a mastoparan dose-dependent increase in synaptic vesicle swelling (Fig. 30f). Twenty micromolar mastoparan (Mas 7) induces a time-dependent (s) increase in synaptic vesicle size (Fig. 30g), as opposed to the control peptide Mast-17. Results from these studies demonstrate that following stimulation of cell secretion, ZGs, the membrane-bound secretory vesicles in exocrine pancreas swell. Different ZGs swell differently, and the extent of their swelling dictates the amount of intravesicular contents expelled. ZG swelling is therefore a requirement for the precise and regulated expulsion of vesicular contents in the exocrine pancreas. Similar to ZGs, synaptic vesicles swell to enable the release of neurotransmitters at the nerve terminal. This mechanism of intravesicular discharge during cell secretion may explain why partially empty vesicles accumulate in secretory cells [105–108] following secretion. The presence of empty secretory vesicles could result from multiple cycles of fusion–swelling–expulsion–dissociation, a vesicle undergoes

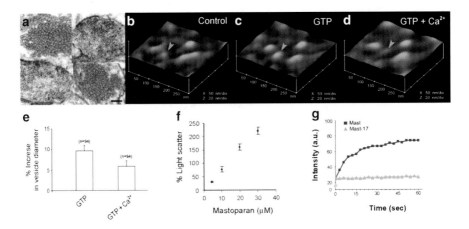

Fig. 30 Synaptic vesicles swell in response to GTP and mastoparan, and vesicle swelling is required for neurotransmitter release. (**a**) Electron micrographs of brain synaptosomes, demonstrating the presence of 40–50-nm synaptic vesicles within. Bar = 200 nm. (**b**) AFM micrographs of synaptosomal membrane, demonstrating the presence of 40–50-nm synaptic vesicles docked to the cytosolic face of the presynaptic membrane. (**c**) Exposure of the synaptic vesicles to 20 μM GTP results in vesicle swelling (*blue arrowhead*). (**d, e**) Furthermore, addition of calcium results in the transient fusion of the synaptic vesicles at porosomes in the presynaptic membrane of the nerve terminal and expulsion of intravesicular contents. Note the decrease in size of the synaptic vesicle following content expulsion. (**f**) Light scattering assays on isolated synaptic vesicles demonstrates the mastoparan dose-dependent increase in vesicle swelling (*n* = 5), and further confirms the AFM results. (**g**) Exposure of isolated synaptic vesicles to 20 μM mastoparan demonstrates a time-dependent (in seconds) increase in their swelling. Note the control peptide mast-17 has little or no effect on synaptic vesicle swelling (Figure from our earlier publication: *Cell Biol Int* 2004, 28:709–716). ©Bhanu Jena

during the secretory process. These results reflect the precise and regulated nature of secretion, both in the exocrine pancreas and in neurons.

Molecular Mechanism of Secretory Vesicle Swelling

Our understanding of the molecular mechanism of secretory vesicle swelling has greatly advanced in the past 15 years. Isolated secretory vesicles, single vesicle patch, and reconstituted swelling-competent proteoliposomes have been utilized [69, 70, 100–104, 109] to determine the mechanism and regulation of vesicle swelling. Isolated ZGs from the exocrine pancreas swell rapidly in response to GTP [69, 70], suggesting rapid water gating into ZGs. Results from studies demonstrate the presence of the water channel aquaporin-1 (AQP1) at the ZG membrane [70] and aquaporin-6 (AQP6) at the synaptic vesicle membrane [100] and their participation in GTP-mediated water entry and vesicle swelling. Furthermore, the molecular regulation of AQP1 at the ZG membrane has been studied [101], providing a general mechanism of secretory vesicle swelling. Detergent-solubilized ZGs immunoiso-

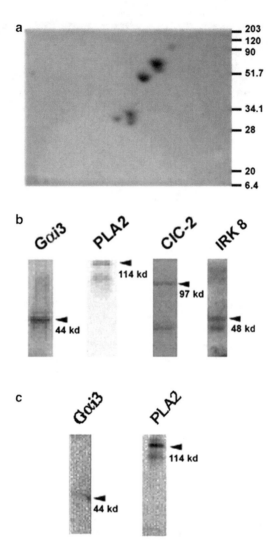

Fig. 31 One (1D) and two dimensionally (2D) resolved, AQP1-immunoisolated proteins from solubilized pancreatic ZG preparations (**a**, **b**) and red blood cells (**c**), using an AQP1-specific antibody. The 2D-resolved proteins were coumassie stained and the 1D-resolved proteins transferred to nitrocellulose membranes for immunoblot analysis. Note the identification in the immunoisolates of seven spots in the coumassie-stained 2D-resolved gel (**a**). Immunoblot analysis of the 1D-resolved immunoisolated proteins demonstrates the presence of PLA$_2$, G$_{a13}$, the potassium channel IRK-8, and chloride channel ClC-2 (**b**). Similarly, in red blood cells, PLA$_2$ and G$_{a13}$ immunoreactive bands are detected (**c**). Lower molecular weight bands may represent proteolytic cleavage products (Figure from our earlier publication: *Cell Biol Int* 2004, 28:7–17). ©Bhanu Jena

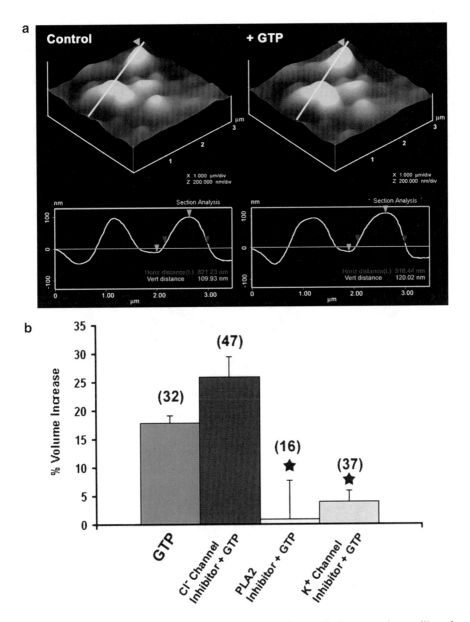

Fig. 32 ZG volume changes measured using AFM. AFM micrographs demonstrating swelling of isolated ZGs in response to 40 μM GTP (**a**). Histogram showing changes in ZG volume following exposure of isolated ZGs to GTP, and to the inhibitor of the chloride channel, DIDS, to the inhibitor of the potassium channel, glyburide, or to the inhibitor of PLA_2, ONO-RS-082 (**b**). Note the significant inhibition of GTP-induced ZG swelling in the presence of glyburide or ONO-RS-082. DIDS has little effect on GTP-induced ZG swelling. Values represent mean ± SE of mean of the number of ZGs (in *parenthesis*), which were randomly selected. A significant ($p < 0.01$) inhibition (*) in the presence of glyburide or ONO-RS-082 was demonstrated (Figure from our earlier publication: *Cell Biol Int* 2004, 28:7–17). ©Bhanu Jena

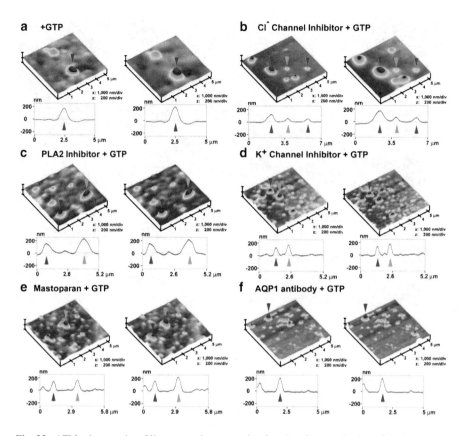

Fig. 33 AFM micrographs of liposomes demonstrating functional reconstitution of the immunoisolated AQP1 complex. Note swelling of the reconstituted liposomes in response to GTP (**a**) *panel left* is control and *panel right* is 5 min following GTP exposure. Exposure of the chloride channel inhibitor DIDS did not inhibit GTP-induced swelling (**b**); however, the PLA$_2$ inhibitor ONO-RS-082 (**c**), the potassium channel inhibitor glyburide (**d**), and the inhibitory AQP1-specific antibody inhibited GTP-induced swelling (**f**). The G$_{\alpha i}$-stimulable peptide mastoparan demonstrates no additional effect on swelling over GTP (**e**); hence, the GTP effect is mediated through the G$_{\alpha i3}$ protein (Figure from our earlier publication: *Cell Biol Int* 2004, 28:7–17). ©Bhanu Jena

lated using monoclonal AQP-1 antibody co-isolates AQP1, PLA2, G$_{\alpha i3}$, potassium channel IRK-8, and the chloride channel ClC-2 ([101], Figs. 31–34). Exposure of ZGs to either the potassium channel blocker glyburide or the PLA2 inhibitor ONO-RS-082 blocks GTP-induced ZG swelling. Red blood cells known to possess AQP1 at the plasma membrane also swell on exposure to the G$_{\alpha i}$ agonist mastoparan and responds similarly to ONO-RS-082 and glyburide, as do ZGs [101]. Artificial liposomes reconstituted with the AQP1 immunoisolated complex from solubilized ZG preparation also swell in response to GTP. Glyburide or ONO-RS-082 is found to abrogate the GTP effect in reconstituted liposomes. AQP1 immunoisolate-reconstituted planar lipid membrane demonstrates conductance, which is sensitive to

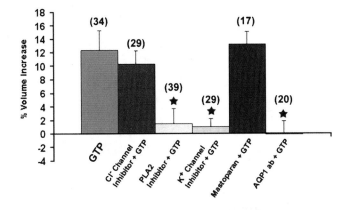

Fig. 34 Percent volume changes in reconstituted liposomes measured by AFM. Estimation of the changes in liposome volume demonstrates a 12–15% increase, 5 min following exposure to 40 μM GTP. No significant change in GTP-induced liposome swelling is observed following exposure to the chloride channel inhibitor DIDS or to the $G_{\alpha i}$-stimulable peptide mastoparan. Similar to the responses in ZGs, exposure of the reconstituted liposomes to the PLA_2 inhibitor ONO-RS-082, the potassium channel inhibitor glyburide, and the AQP1-specific antibody significantly inhibits the GTP effect. Values represent mean ± SE of mean of randomly selected liposomes (in *parenthesis*). A significant ($p < 0.01$) inhibition (*) in the presence of glyburide, ONO-RS-082, and the AQP1-specific antibody, is demonstrated (Figure from our earlier publication: *Cell Biol Int* 2004, 28:7–17). ©Bhanu Jena

glyburide and an AQP1-specific antibody. These results demonstrate a Gαi3-PLA2 mediated pathway and potassium channel involvement in AQP1 regulation at the ZG membrane [101], contributing to ZG swelling. Similarly, AQP-6 involvement has been demonstrated in GTP-induced and G_o-mediated synaptic vesicle swelling in neurons [100].

To further characterize the ion channels present at the secretory vesicle membrane, for the first time studies were carried out using single ZG patch [109]. These studies confirm earlier findings of the presence of both potassium and chloride ion channels at the ZG membrane. In these studies, the electrical activity at the ZG membrane displays a range of sensitivity both to chloride and to potassium channel blockers. Whole vesicle conductance was decreased with the addition of the chloride channel blocker, DIDS, and the ATP K + channel blocker, glyburide, in both vesicles patches and indirect analysis, supporting the hypothesis for the presence of more than one channel type [109]. This finding was further confirmed immunochemically using western blot analysis, and as speculated, the presence of two chloride channels, CLC-2 and CLC-3, was observed [109]. Also consistent with pharmacological evidence was the presence of ATP-sensitive potassium channel, Kir6.1, in western blot analysis of the ZGs. This is surprising since Kir6.2 is the predominant form of potassium channels in β-cells of the endocrine pancreas [109].

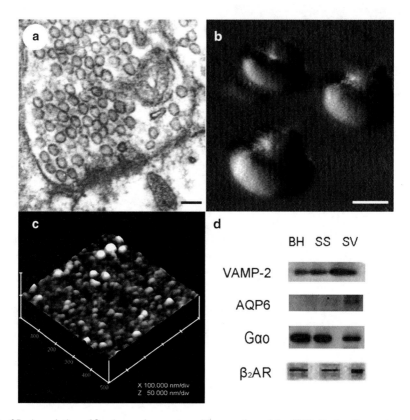

Fig. 35 Association of β₂ adrenergic receptor with synaptic vesicles (SV). Purity of synaptosomes (**a, b**) and SV (**c**) was determined using transmission electron microscopy (**a**, bar = 100 nm), atomic force microscopy (**b**, bar = 1 μm, and **c**), and immunoblot analysis (**d**) on total rat brain homogenate (BH), synaptosome (SS), and SV (SV). Note the clean 1–2-μm mushroom-shaped synatosomes (**b**) and the 35–50-nm SV (**c**) preparations in the atomic force micrographs. Immunoblot analysis of 10 μg protein each of BH, SS, and SV demonstrates the presence of G$_{\alpha o}$ protein, and the enriched presence of SV proteins VAMP-2 and the water channel AQP6. Note the enriched presence of β₂AR in the SV fraction (Figure from our earlier publication: *J Cell Mol Med* 2010, 15:572–576). ©Bhanu Jena

Presence of Functional β-Adrenergic Receptors at the Synaptic Vesicle Membrane

Since mastoparan, an amphiphilic tetradecapeptide from wasp venom, activates G$_o$ protein GTPase and stimulates synaptic vesicle swelling, the presence of β-adrenergic receptor at the synaptic vesicle membrane was hypothesized. Stimulation of G proteins is believed to occur via insertion of mastoparan into the phospholipid membrane to form a highly structured α-helix that resembles the intracellular loops

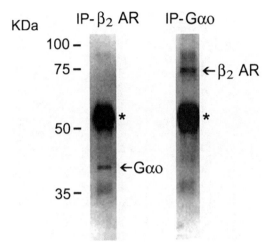

Fig. 36 Interaction of β_2 adrenergic receptor with $G_{\alpha o}$ protein in synaptic vesicles (SV). Detergent-solubilized SV preparation immunoprecipitated using β_2AR-specific antibody results in co-isolation of the 45-kDa $G_{\alpha o}$ protein, suggesting their physical association in SV. Similarly, when detergent-solubilized SV preparation was immunoprecipitated using $G_{\alpha o}$-specific antibody and the resultant isolate was then probed with the β_2AR-specific antibody, the 68-kDa β_2AR protein is co-isolated. The *dark band* represented by (*) is the heavy chain of the antibody used in immunoprecipitation reaction (Figure from our earlier publication: *J Cell Mol Med* 2010, 15:572–576). ©Bhanu Jena

of G protein-coupled adrenergic receptors. Consequently, the presence of adreno-ceptors and of an endogenous β-adrenergic agonist at the synaptic vesicle membrane has been investigated. Immunoblot analysis of synaptic vesicle using β-adrenergic receptor antibody (Figs. 34 and 35), and vesicle swelling experiments using β-adrenergic agonists and antagonists, demonstrate the presence of functional β-adrenergic receptors at the synaptic vesicle membrane (Figs. 37 and 38, [104]). Electron microscopy (EM) (Fig. 35a), AFM (Fig. 35b, c), and immunoblot analysis (Fig. 35d) demonstrated a highly enriched synaptosome (SS) and synaptic vesicle (SV) preparation. Immunoblot analysis demonstrates the SV preparation to be enriched in VAMP-2 and AQP-6 (Fig. 35d), both SV-specific proteins [100]. Additionally, in conformation with earlier findings [100], the GTP-binding $G_{\alpha o}$ protein is found to associate with the SV preparation. Collectively, these studies demonstrate the isolation of a highly enriched SV preparation from rat brain tissue, for SV swelling and β_2AR immunolocalization assays. To determine the relative concentration of β_2AR in SV, immunoblot analysis was performed using 10 μg each of total brain homogenate (BH), isolated synaptosome (SS), and synaptic vesicle (SV) fractions (Fig. 34). In conformation with earlier findings [110], β_2AR was present both in the BH and in the SS fraction. However, for the first time, and in conformation with our hypothesis, β_2AR was demonstrated to be present and enriched in the SV fraction (Fig. 35d).

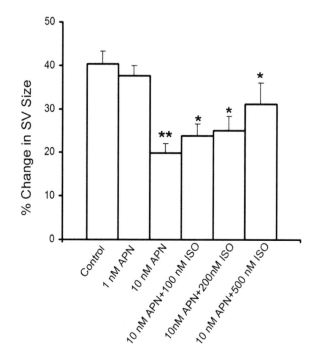

Fig. 37 β_2 adrenergic receptors in synaptic vesicles (SV) are functional. Photon correlation spectroscopy (PCS) used in the determination of SV size. Exposure of SV to 40 µM GTP and 40 µM mastoparan results in a significant increase in SV size (control). However, exposure of SV to the β_2AR antagonist alprenolol (APN) demonstrates inhibition of the GTP–mastoparan-mediated SV swelling. Note the significant ($n=6$, **$p<0.001$) inhibition of GTP–mastoparan-mediated SV swelling in the presence of 10 nM APN. Exposure of the 10 nM APN-treated SV to the β_2AR agonist isoproterenol (ISO) demonstrated a dose-dependent increase ($n=6$, *$p<0.05$) in SV size (Figure from our earlier publication: *J Cell Mol Med* 2010, 15:572–576). ©Bhanu Jena

Upon binding to endogenous activators, adrenergic receptors undergo a conformational change that leads to the activation of heterotrimeric GTP-binding proteins [13]. Different groups of adrenergic receptors couple to and activate only certain G protein types, thus leading to specific intracellular signals [111]. Our immunoblot assays demonstrate the enriched presence of β2 adrenergic receptors in the SV fraction; therefore, the physical interaction of SV-associated $G_{\alpha o}$ and β2 adrenergic receptors was investigated. To determine the co-association of $G_{\alpha o}$ and β_2AR at the SV, immunopulldown and immunoblot analyses were carried out. Detergent-solubilized SV preparations were immunoprecipitated using β_2AR-specific antibody and the resultant isolate was probed with a $G_{\alpha o}$-specific primary antibody. Results from the study demonstrate the co-isolation of a 45-kDa $G_{\alpha o}$-specific protein, suggesting the physical association of $G_{\alpha o}$ and the β_2AR in SV (Fig. 36). To further confirm this interaction, the detergent-solubilized SV preparation was once again immunoprecipitated, this time using $G_{\alpha o}$-specific antibody and the resultant isolate probed with the β_2AR-specific antibody. Once again in agreement, our results demonstrate

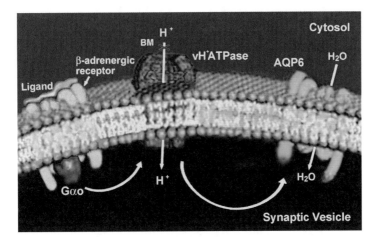

Fig. 38 Schematic diagram of synaptic vesicle membrane, depicting the presence and participation of β_2AR-$G_{\alpha o}$, vH+-ATPase, and the water channel AQP6, is vesicle swelling. The current study shows that GTP-binding $G_{\alpha o}$ protein is stimulated by the activated β_2AR via its endogenous ligand. Earlier studies demonstrate the involvement of vH+-ATPase in GTP–$G_{\alpha o}$-mediated synaptic vesicle swelling. Bafilomycin (BM)-sensitive (*red X*) vesicle acidification following the GTP-$G_{\alpha o}$ stimulus has also been demonstrated, and since water channels are bidirectional and the vH+-ATPase inhibitor BM decreases the volume of both isolated SV- and GTP–mastoparan-stimulated swelling, suggests vH+-ATPase to be upstream of AQP-6. Therefore, the SV swelling pathway involves occupancy of the β_2AR and, as a consequence, the activation of the GTP-binding $G_{\alpha o}$ protein followed by activation of vH+-ATPase and transport of water via AQP6 into SV (Figure from our earlier publication: *J Cell Mol Med* 2010, 15:572–576). ©Bhanu Jena

the co-isolation of the 68-kDa β_2AR and $G_{\alpha o}$, suggesting the physical association of $G_{\alpha o}$ and β_2AR in SV (Fig. 36).

The functional state of the β_2AR in SV and its role in regulating SV swelling were determined using photon correlation spectroscopy (PCS) in the presence and absence of both β_2AR agonist (isoproterenol) and antagonist (alprenolol). Conforming with earlier studies [100], exposure of SV to 40 µM GTP and 40 µM mastoparan resulted in a significant increase in SV size. When SV preparations were exposed to various concentrations of the β_2 adrenergic receptor antagonist alprenolol, GTP- and mastoparan-induced swelling was decreased in a dose-dependent manner. Ten nanomolar alprenolol significantly decreases the GTP/mastoparan-induced SV swelling (Fig. 37). To determine the effect of β_2 adrenergic receptor agonist isoproterenol, increasing concentrations (100, 200, and 500 nM) of the ligand were used following alprenolol inhibition. Results from this study demonstrate that all three concentrations of the agonist significantly stimulate GTP/mastoparan-induced SV swelling, with 500 nM isoproterenol having the greatest effect (Fig. 37). Pre-exposure of SV to alprenolol resulted in isoproterenol stimulation of GTP/mastoparan-induced SV swelling (Fig. 37), suggesting that perhaps β_2 adrenergic receptors are already preoccupied by some endogenous ligand, and the receptor–ligand complex may be present in a desensitized state at the SV membrane. Therefore, following exposure to alprenolol, the antagonist is able to dislodge this endogenous ligand from the receptor, enabling the agonist isoproterenol to then replace the antagonist alprenolol to bind and stimulate the SV-associated β_2 adrenergic receptor, causing SV swelling.

These studies demonstrate for the first time that functional β_2 adrenergic receptors, and endogenous ligands, are associated with SV, regulating vesicle swelling in neurons. Since vH+-ATPase activity is upstream of AQP-6 in the pathway leading from $G_{\alpha o}$-stimulated swelling of SV, binding of the endogenous β-adrenergic agonist to its receptors at the SV membrane likely initiates the swelling cascade (Fig. 38).

Conclusion

In this chapter, the current understanding of the cellular and molecular physiology of secretion is presented. Porosomes are specialized plasma membrane structures universally present in secretory cells, from exocrine and endocrine cells to neuroendocrine cells and neurons. Since porosomes in exocrine and neuroendocrine cells measure 100–180 nm, and only a 20–35% increase in porosome diameter is demonstrated following the docking and fusion of 0.2–1.2 µm in diameter secretory vesicles, it is concluded that secretory vesicles "transiently" dock and fuse at the base of the porosome complex to release their contents to the outside. Furthermore, isolated live cells in near physiological buffer when imaged using AFM demonstrate the size and shape of the secretory vesicles lying immediately below the apical plasma membrane of the cell. Following exposure to a secretory stimulus secretory vesicles swell, followed by a decrease in vesicle size. No loss of secretory vesicles is observed following secretion, demonstrating transient fusion and partial discharge of vesicular contents during cell secretion. In agreement, studies demonstrate that "secretory granules are recaptured largely intact after stimulated exocytosis in cultured endocrine cells" [44], "single synaptic vesicles fuse transiently and successively without loss of identity" [45], "zymogen granule exocytosis is characterized by long fusion pore openings and preservation of vesicle lipid identity" [46]. This is in contrast to the general belief that in mammalian cells, secretory vesicles completely merge at the cell plasma membrane, resulting in passive diffusion of vesicular contents to the cell exterior, and the consequent retrieval of excess membrane by endocytosis at a later time. Additionally, a major logistical problem with complete merger of secretory vesicle membrane at the cell plasma membrane is the generation of partially empty vesicles following cell secretion. It is fascinating how even single-cell organisms have developed such specialized secretory machinery as the secretion apparatus of *Toxoplasma gondii*, the contractile vacuole in paramecium, and the secretory structures in bacteria. Hence, it comes as no surprise that mammalian cells have evolved such highly specialized and sophisticated structures—the porosome complex for cell secretion. The discovery of the porosome, and an understanding of its structure and dynamics at nanometer resolution and in real time in live cells, its composition, and its functional reconstitution in lipid membrane, and the molecular mechanism of SNARE-induced membrane fusion, has greatly advanced our understanding of cell secretion. It is evident that the secretory process in cells is a well coordinated, highly regulated, and a finely tuned biomolecular orchestra.

Results of studies discussed in this chapter further demonstrate that in the presence of Ca^{2+}, t-SNAREs and v-SNARE in opposing bilayers interact and self-assemble in a circular pattern to form conducting channels. Such self-assembly of t-/v-SNAREs in a ring conformation occurs only when the respective SNAREs are in association with membrane. X-ray diffraction measurements further demonstrate that t-SNAREs in the target membrane and v-SNARE in the vesicle membrane overcome repulsive forces to bring opposing membranes close to within a distance of 2.8 Å. Studies provide evidence that calcium bridging of the opposing bilayers leads to release of water from hydrated Ca^{2+} ions, as well as the loosely coordinated water at PO-lipid head groups, leading to membrane destabilization and fusion. The t-/v-SNARE is a tight complex whose disassembly requires an ATPase called NSF, which functions as a right-handed molecular motor.

Similarly in vivo and in vitro measurements of secretory vesicle dynamics demonstrate that vesicle swelling is required for the expulsion of intravesicular content from cells during secretion. It is demonstrated that the amount of intravesicular contents expelled is directly proportional to the extent of secretory vesicle swelling. This unique capability provides cells with the ability to precisely regulate the release of secretory products during cell secretion. The direct observation in live cells using the AFM, and the requirement of secretory vesicle swelling in cell secretion, further explains the appearance of empty and partially empty vesicles following cell secretion. In recent years, there has been a flurry of research in the field, and a number of papers from several laboratories investigating the porosome in different cell types, both in their native and in disease states [112–117], such as porosome in the sensory hair cell and in RBL-2H3 and BMMC cells. Clearly, these findings could not have advanced without the AFM, and therefore this powerful tool has greatly contributed to a new understanding of the cell. The AFM has enabled the determination of live cellular structure–function at subnanometer to angstrom resolution, in real time, contributing to the birth of the new field of *NanoCellBiology*. Future directions will involve an understanding of the protein distribution and their arrangement at atomic resolution in the porosome complex, as well as an understanding of the atomic structure of SNAREs within the membrane-associated t-/v-SNARE ring complex. Currently, these goals are being pursued using electron crystallography.

Acknowledgments The author thanks the students and collaborators who have participated in the various studies discussed in this chapter. Support from the National Institutes of Health (USA), the National Science Foundation (USA), and Wayne State University is greatly appreciated.

References

1. Katz B (1962) The transmission of impulses from nerve to muscle and the subcellular unit of synaptic action. Proc R Soc Lond B Biol Sci 155:455–479
2. Folkow B, Häggendal J, Lisander B (1967) Extent of release and elimination of noradrenalin at peripheral adrenergic nerve terminal. Acta Physiol Scand Suppl 307:1–38
3. Folkow B, Häggendal J (1970) Some aspects of the quantal release of the adrenergic transmitter. Springer-Verlag Bayer Symp II:91–97

4. Folkow B (1997) Transmitter release at the adrenergic nerve endings: total exocytosis or fractional release? News Physiol Sci 12:32–35
5. Neher E (1993) Secretion without full fusion. Nature 363:497–498
6. Schneider SW, Sritharan KC, Geibel JP, Oberleithner H, Jena BP (1997) Surface dynamics in living acinar cells imaged by atomic force microscopy: identification of plasma membrane structures involved in exocytosis. Proc Natl Acad Sci U S A 94:316–321
7. Cho S-J, Quinn AS, Stromer MH, Dash S, Cho J, Taatjes DJ, Jena BP (2002) Structure and dynamics of the fusion pore in live cells. Cell Biol Int 26:35–42
8. Cho S-J, Wakade A, Pappas GD, Jena BP (2002) New structure involved in transient membrane fusion and exocytosis. Ann New York Acad Sci 971:254–256
9. Cho S-J, Jeftinija K, Glavaski A, Jeftinija S, Jena BP, Anderson LL (2002) Structure and dynamics of the fusion pores in live GH-secreting cells revealed using atomic force microscopy. Endocrinology 143:1144–1148
10. Cho WJ, Jeremic A, Rognlien KT, Zhvania MG, Lazrishvili I, Tamar B, Jena BP (2004) Structure, isolation, composition and reconstitution of the neuronal fusion pore. Cell Biol Int 28:699–708
11. Cho WJ, Jeremic A, Jin H, Ren G, Jena BP (2007) Neuronal fusion pore assembly requires membrane cholesterol. Cell Biol Int 31:1301–1308
12. Cho WJ, Ren G, Jena BP (2008) EM 3D contour maps provide protein assembly at the nanoscale within the neuronal porosome complex. J Microsc 232:106–111
13. Jena BP, Cho S-J, Jeremic A, Stromer MH, Abu-Hamdah R (2003) Structure and composition of the fusion pore. Biophys J 84:1–7
14. Jeremic A, Kelly M, Cho S-J, Stromer MH, Jena BP (2003) Reconstituted fusion pore. Biophys J 85:2035–2043
15. Jena BP (2011) Porosome: the universal secretory portal in cells. Biomed Rev 21:1–15
16. Jena BP (2010) Secretory vesicles transiently dock and fuse at the porosome to discharge contents during cell secretion. Cell Biol Int 34:3–12
17. Jena BP (2009) Functional organization of the porosome complex and associated structures facilitating cellular secretion. Physiology 24:367–376
18. Jena BP (2009) Porosome: the secretory portal in cells. Biochemistry 49:4009–4018
19. Jena BP (2008) Porosome: the universal molecular machinery for cell secretion. Mol Cells 26:517–529
20. Jena BP (2007) Secretion machinery at the cell plasma membrane. Curr Opin Struct Biol 17:437–443
21. Jena BP (2006) Cell secretion machinery: studies using the AFM. Ultramicroscopy 106:663–669
22. Jena BP (2005) Cell secretion and membrane fusion. Domest Anim Endocrinol 29:145–165
23. Jena BP (2005) Molecular machinery and mechanism of cell secretion. Exp Biol Med 230:307–319
24. Jena BP (2004) Discovery of the porosome: revealing the molecular mechanism of secretion and membrane fusion in cells. J Cell Mol Med 8:1–21
25. Jena BP (2003) Fusion pore: structure and dynamics. J Endocrinol 176:169–174
26. Jena BP (2002) Fusion pores in live cells. News Physiol Sci 17:219–222
27. Holden C (1997) Early peek at a cellular porthole. Science 275:485
28. Wheatley DN (2004) A new frontier in cell biology: nano cell biology. Cell Biol Int 28:1–2
29. Anderson LL (2004) Discovery of a new cellular structure—the porosome: elucidation of the molecular mechanism of secretion. Cell Biol Int 28:3–5
30. Singer MV (2004) Legacy of a distinguished scientist: George E. Palade. Pancreatology 3:518–519
31. Craciun C (2004) Elucidation of cell secretion: pancreas led the way. Pancreatology 4:487–489
32. Anderson LL (2006) Discovery of the 'porosome'; the universal secretory machinery in cells. J Cell Mol Med 10:126–131
33. Hörber JKH, Miles MJ (2003) Scanning probe evolution in biology. Science 302:1002–1005
34. Allison DP, Doktyez MJ (2006) Cell secretion studies by force microscopy. J Cell Mol Med 10:847–856

35. Anderson LL (2006) Cell secretion finally sees the light. J Cell Mol Med 10:270–272
36. Jeftinija S (2006) The story of cell secretion: events leading to the discovery of the 'porosome'—the universal secretory machinery in cells. J Cell Mol Med 10:273–279
37. Jeremic A (2008) Cell secretion: an update. J Cell Mol Med 12:1151–1154
38. Labhasetwar V (2007) A milestone in science: discovery of the porosome—the universal secretory machinery in cells. J Biomed Nanotechnol 3:1
39. Leabu M (2006) Discovery of the molecular machinery and mechanisms of membrane fusion in cells. J Cell Mol Med 10:423–427
40. Wheatley DN (2007) Pores for thought: further landmarks in the elucidation of the mechanism of secretion. Cell Biol Int 31:1297–1300
41. Paknikar KM, Jeremic A (2007) Discovery of the cell secretion machinery. J Biomed Nanotechnol 3:218–222
42. Paknikar KM (2007) Landmark discoveries in intracellular transport and secretion. J Cell Mol Med 11:393–397
43. Siksou L, Rostaing P, Lechaire JP, Boudier T, Ohtsuka T, Fejtova A, Kao HT, Greengard P, Gundelfinger ED, Triller A, Marty S (2007) Three-dimensional architecture of presynaptic terminal cytomatrix. J Neurosci 27:6868–6877
44. Taraska JW, Perrais D, Ohara-Imaizumi M, Nagamatsu S, Almers W (2003) Secretory granules are recaptured largely intact after stimulated exocytosis in cultured endocrine cells. Proc Natl Acad Sci U S A 100:2070–2075
45. Aravanis AM, Pyle JL, Tsien RW (2003) Single synaptic vesicles fusing transiently and successively without loss of identity. Nature 423:643–647
46. Thorn P, Fogarty KE, Parker I (2004) Zymogen granule exocytosis is characterized by long fusion pore openings and preservation of vesicle lipid identity. Proc Natl Acad Sci U S A 101:6774–6779
47. Kuznetsov SA, Langford GM, Weiss DG (1992) Actin-dependent organelle movement in squid axoplasm. Nature 356:722–725
48. Schroer TA, Sheetz MP (1991) Functions of microtubule-based motors. Annu Rev Physiol 53:629–652
49. Evans LL, Lee AJ, Bridgman PC, Mooseker MS (1998) Vesicle-associated brain myosin-V can be activated to catalyze actin-based transport. J Cell Sci 111:2055–2066
50. Rudolf R, Kogel T, Kuznetsov SA, Salm T, Sclicker O, Hellwig A, Hammer JA, Gerdes H-H (2003) Myosin Va facilitates the distribution of secretory granules in the F-actin rich corted of PC12 cells. J Cell Sci 116:1339–12348
51. Varadi A, Tsuboi T, Rutter GA (2005) Myosin Va transports dense core secretory vesicles in pancreatic MIN6 beta-cells. Mol Biol Cell 16:2670–2680
52. Cheney RE, O'Shea MK, Heuser JE, Coelho MV, Wolenski JS, Espreafico EM, Forscher P, Larson RE, Mooseker MS (1993) Brain myosin-V is a two-headed unconventional myosin with motor activity. Cell 75:13–23
53. Reck-Peterson SL, Provance DW Jr, Mooseker MS, Mercer JA (2000) Class V myosins. Biochim Biophys Acta 1496:36–51
54. Hirschberg K, Miller CM, Ellenberg J, Presley JF, Siggia ED, Phair RD, Lippincott-Schwartz J (1998) Kinetic analysis of secretory protein traffic and characterization of golgi to plasma membrane transport intermediates in living cells. J Cell Biol 143:1485–1503
55. Rudolf R, Salm T, Rustom A, Gerdes H-H (2001) Dynamics of immature secretory granules: role of cytoskeletal elements during transport, cortical restriction, and F-actin-dependent tethering. Mol Biol Cell 12:1353–1365
56. Manneville J-B, Etienne-Manneville S, Skehel P, Carter T, Ogden D, Ferenczi M (2003) Interaction of the actin cytoskeleton with microtubules regulates secretory organelle movement near the plasma membrane in human endothelial cells. J Cell Sci 116:3927–3938
57. Abu-Hamdah R, Cho W-J, Hörber JKH, Jena BP (2006) Secretory vesicles in live cells are not free-floating but tethered to filamentous structures: a study using photonic force microscopy. Ultramicroscopy 106:670–673
58. Jeremic A, Kelly M, Cho J-H, Cho S-J, Horber JKH, Jena BP (2004) Calcium drives fusion of SNARE-apposed bilayers. Cell Biol Int 28:19–31

59. Malhotra V, Orci L, Glick BS, Block MR, Rothman JE (1988) Role of an *N*-ethylmaleimide-sensitive transport component in promoting fusion of transport vesicles with cisternae of the Golgi stack. Cell 54:221–227

60. Trimble WS, Cowan DW, Scheller RH (1988) VAMP-1: A synaptic vesicle-associated integral membrane protein. Proc Natl Acad Sci U S A 85:4538–4542

61. Oyler GA, Higgins GA, Hart RA, Battenberg E, Billingsley M, Bloom FE, Wilson MC (1989) The identification of a novel synaptosomal-associated protein, SNAP-25, differentially expressed by neuronal subpopulations. J Cell Biol 109:3039–3052

62. Bennett MK, Calakos N, Schller RH (1992) Syntaxin: A synaptic protein implicated in docking of synaptic vesicles at presynaptic active zones. Science 257:255–259

63. Cho S-J, Kelly M, Rognlien KT, Cho J, Hörber JK, Jena BP (2002) SNAREs in opposing bilayers interact in a circular array to form conducting pores. Biophys J 83:2522–2527

64. Cho WJ, Jeremic A, Jena BP (2005) Size of supramolecular SNARE complex: membrane-directed self-assembly. J Am Chem Soc 127:10156–10157

65. Cho WJ, Lee J-S, Ren G, Zhang L, Shin L, Manke CW, Potoff J, Kotaria N, Zhvania MG, Jena BP (2011) Membrane-directed molecular assembly of the neuronal SNARE complex. J Cell Mol Med 15:31–37

66. Jeremic A, Cho WJ, Jena BP (2004) Membrane fusion: what may transpire at the atomic level. J Biol Phys Chem 4:139–142

67. Potoff JJ, Issa Z, Manke CW Jr, Jena BP (2008) Ca^{2+}-dimethylphosphate complex formation: providing insight into Ca^{2+} mediated local dehydration and membrane fusion in cells. Cell Biol Int 32:361–366

68. Kelly M, Cho WJ, Jeremic A, Abu-Hamdah R, Jena BP (2004) Vesicle swelling regulates content expulsion during secretion. Cell Biol Int 28:709–716

69. Jena BP, Schneider SW, Geibel JP, Webster P, Oberleithner H, Sritharan KC (1997) G_i regulation of secretory vesicle swelling examined by atomic force microscopy. Proc Natl Acad Sci U S A 94:13317–13322

70. Cho S-J, Sattar AK, Jeong EH, Satchi M, Cho J, Dash S, Mayes MS, Stromer MH, Jena BP (2002) Aquaporin 1 regulates GTP-induced rapid gating of water in secretory vesicles. Proc Natl Acad Sci U S A 99:4720–4724

71. Binnig G, Quate CF, Gerber CH (1986) Atomic force microscope. Phys Rev Lett 56:930–933

72. Alexander S, Hellemans L, Marti O, Schneir J, Elings V, Hansma PK (1989) An atomic resolution atomic force microscope implemented using an optical lever. J Appl Phys 65:164–167

73. Gaisano HY, Sheu L, Wong PP, Klip A, Trimble WS (1997) SNAP-23 is located in the baso-lateral plasma membrane of rat pancreatic acinar cells. FEBS Lett 414:298–302

74. Lee J-S, Cho W-J, Jeftinija K, Jeftinija S, Jena BP (2009) Porosome in astrocytes. J Cell Mol Med 13:365–372

75. Cho W-J, Shin L, Ren G, Jena BP (2009) Structure of membrane-associated neuronal SNARE complex: Implication in neurotransmitter release. J Cell Mol Med 13:4161–4165

76. Mohrmann R, de Wit H, Verhage M, Neher E, Sørensen JB (2010) Fast vesicle fusion in living cells requires at least three SNARE complexes. Science 330:502–505

77. Ludtke SJ, Baldwin PR, Chiu W (1999) EMAN: semiautomated software for high-resolution single-particle reconstructions. J Struct Biol 128:82–97

78. Frank J, Radermacher M, Penczek P, Zhu J, Li Y, Lasjadj M, Leith A (1996) SPIDER and WEB: processing and visualization of images in 3D electron microscopy and related fields. J Struct Biol 116:190–199

79. Goddard TD, Huang CC, Ferrin TE (2005) Software extensions to UCSF Chimera for inter-active visualization of large molecular assemblies. Structure 13:473–482

80. Pettersen EF, Goddard TD, Huang CC, Couch GS, Greenblatt DM, Meng EC, Ferrin TE (2004) UCSF chimera—a visualization system for exploratory research and analysis. J Comput Chem 25:1605–1612

81. Cho WJ, Jeremic A, Jena BP (2005) Direct interaction between SNAP-23 and L-type calcium channel. J Cell Mol Med 9:380–386

82. Cho WJ, Jena BP (2007) N-ethymaleimide sensitive factor is a right-handed molecular motor. J Biomed Nanotechnol 3:209–211

83. Jeremic A, Quinn AS, Cho WJ, Taatjes DJ, Jena BP (2006) Energy-dependent disassembly of self-assembled SNARE complex: observation at nanometer resolution using atomic force microscopy. J Am Chem Soc 128:26–27

84. Weber T, Zemelman BV, McNew JA, Westerman B, Gmachi M, Parlati F, Söllner TH, Rothman JE (1998) SNAREpins: minimal machinery for membrane fusion. Cell 92:759–772

85. Bako I, Hutter J, Palinkas G (2002) Car-Parrinello molecular dynamics simulation of the hydrated calcium ion. J Chem Phys 117:9838–9843

86. McIntosh TJ (2000) Short-range interactions between lipid bilayers measured by X-ray diffraction. Curr Opin Struct Biol 10:481–485

87. Portis A, Newton C, Pangborn W, Papahadjopoulos D (1979) Studies on the mechanism of membrane fusion: evidence for an intermembrane Ca^{2+} phospholipid complex, synergism with Mg^{2+}, and inhibition by spectrin. Biochemistry 18:780–790

88. Kelly ML, Woodbury DJ (1996) Ion channels from synaptic vesicle membrane fragments reconstituted into lipid bilayers. Biophys J 70:2593–2599

89. Cook JD, Cho WJ, Stemmler TL, Jena BP (2008) Circular dichroism (CD) spectroscopy of the assembly and disassembly of SNAREs: the proteins involved in membrane fusion in cells. Chem Phys Lett 462:6–9

90. Woodbury DJ, Miller C (1990) Nystatin-induced liposome fusion. A versatile approach to ion channel reconstitution into planar bilayers. Biophys J 58:833–839

91. Woodbury DJ (1999) Nystatin/ergosterol method for reconstituting ion channels into planar lipid bilayers. Methods Enzymol 294:319–339

92. Cohen FS, Niles WD (1993) Reconstituting channels into planar membranes: a conceptual framework and methods for fusing vesicles to planar bilayer phospholipid membranes. Methods Enzymol 220:50–68

93. Jeong E-H, Webster P, Khuong CQ, Abdus Sattar AK, Satchi M, Jena BP (1999) The native membrane fusion machinery in cells. Cell Biol Int 22:657–670

94. McMahon HT, Gallop JL (2005) Membrane curvature and mechanisms of dynamic cell membrane remodelling. Nature 438:590–596

95. Chernomordik L (1996) Non-bilayer lipids and biological fusion intermediates. Chem Phys Lipids 81:203–213

96. Shin L, Cho W-J, Cook J, Stemmler T, Jena BP (2010) Membrane lipids influence protein complex assembly-disassembly. J Am Chem Soc 132:5596–5597

97. Mitter D, Reisinger C, Hinz B, Hollmann S, Yelamanchili SV, Treiber-Held S, Ohm TG, Herrmann A, Ahnert-Hilger GJ (2003) The synaptophysin/ synaptobrevin interaction critically depends on the cholesterol content. J Neurochem 84:35–42

98. Shin L, Wang S, Lee J-S, Flack A, Mao G, Jena BP (2011) Phosphatidylcholine inhibits membrane-associated SNARE complex disassembly. J Cell Mol Med. doi:10.1111/j.1582-4934.2011.01433.x

99. Stiasny K, Heinz FX (2004) Differneces in the postfusion confirmations of full-length and truncated class II fusion protein E of tick-borne encephalitis virus. J Virol 78:8536–8542

100. Jeremic A, Cho W-J, Jena BP (2005) Involvement of water channels in synaptic vesicle swelling. Exp Biol Med 230:674–680

101. Abu-Hamdah R, Cho W-J, Cho S-J, Jeremic A, Kelly M, Ilie AE, Jena BP (2004) Regulation of the water channel aquaporin-1: isolation and reconstitution of the regulatory complex. Cell Biol Int 28:7–17

102. Shin L, Basi N, Lee J-S, Cho W-J, Chen Z, Abu-Hamdah R, Oupicky D, Jena BP (2010) Involvement of vH+-ATPase in synaptic vesicle swelling. J Neurosci Res 88:95–101

103. Lee J-S, Cho W-J, Shin L, Jena BP (2010) Involvement of cholesterol in synaptic vesicle swelling. Exp Biol Med 235:470–477

104. Chen Z-H, Lee J-S, Shin L, Cho W-J, Jena BP (2010) Involvement of β-adrenergic receptor in synaptic vesicle swelling and implication in neurotransmitter release. J Cell Mol Med 15:572–576

105. Cho S-J, Jena BP (2002) Number of secretory vesicles remain unchanged following exocytosis. Cell Biol Int 26:29–33

106. Lawson D, Fewtrell C, Gomperts B, Raff M (1975) Anti-immunoglobulin-induced histamine secretion by rat peritoneal mast cells studied by immunoferritin electron microscopy. J Exp Med 142:391–401

107. Plattner H, Atalejo AR, Neher E (1997) Ultrastructural organization of bovine chromaffin cell cortex-analysis by cryofixation and morphometry of aspects pertinent to exocytosis. J Cell Biol 139:1709–1717

108. Lee J-S, Mayes MS, Stromer MH, Scanes CG, Jeftinija S, Anderson LL (2004) Number of Secretory vesicles in growth hormone cells of the pituitary remains unchanged after secretion. Exp Biol Med 229:632–639

109. Kelly M, Abu-Hamdah R, Cho S-J, Ilie AL, Jena BP (2005) Patch clamped single pancreatic zymogen granules: direct measurement of ion channel activities at the granule membrane. Pancreatology 5:443–449

110. Yudowski GA, Puthenveedu MA, von Zastrow M (2006) Distinct modes of regulated receptor insertion to the somatodendritic plasma membrane. Nat Neurosci 9:622–627

111. Wettschureck N, Offermanns S (2005) Mammalian G proteins and their cell type specific functions. Physiol Rev 85:1159–1204

112. Elshennawy WW (2011) Image processing and numerical analysis approaches of porosome in mammalian pancreatic acinar cell. J Am Sci 7:835–843

113. Savigny P, Evans J, McGarth KM (2007) Cell membrane structures during exocytosis. Endocrinology 148:3863–3874

114. Matsuno A, Itoh J, Mizutani A, Takekoshi S, Osamura RY, Okinaga H, Ide F, Miyawaki S, Uno T, Asano S, Tanaka J, Nakaguchi H, Sasaki M, Murakami M (2008) Co-transfection of EYFP-GH and ECFP-rab3B in an experimental pituitary GH3 cell: a role of rab3B in secretion of GH through porosome. Folia Histochem Cytobiol 46:419–421

115. Drescher DG, Cho WJ, Drescher MJ (2011) Identification of the porosome complex in the hair cell. Cell Biol Int Rep 18:31–34

116. Okuneva VG, Japaridze ND, Kotaria NT, Zhvania MG (2011) Neuronal porosome in the rat and cat brain: electron microscopic study. J Tcitologiya (in press)

117. Zhao D, Lulevich V, Liu F, Liu G (2010) Applications of atomic force microscopy in biophysical chemistry. J Phys Chem B 114:5971–5982

Index

B.P. Jena, *NanoCellBiology of Secretion: Imaging Its Cellular and Molecular
Underpinnings*, SpringerBriefs in Biological Imaging, DOI 10.1007/978-1-4614-2438-3,
© Springer Science+Business Media, LLC 2012